IMMUNE RNA

CRC

Immune RNA

Editor:

Edward P. Cohen

Associate Professor
La Rabida-University of Chicago Institute
Departments of Medicine and Microbiology
Division of Biological Sciences
University of Chicago
Chicago, Illinois

A. Arthur Gottlieb
Editor-in-Chief
Immunology Series

Professor and Chairman
Department of Microbiology and Immunology
School of Medicine
Tulane University
New Orleans, Louisiana

Published by

CRC PRESS

CRC PRESS, Inc.
18901 Cranwood Parkway · Cleveland, Ohio 44128

Library of Congress Cataloging in Publication Data

Main entry under title:

Immune RNA.

Bibliography: p.
Includes index.
1. Immune Response. 2. Ribonucleic acid.
I. Cohen, Edward P. [DNLM: 1. RNA – Immunology.
2. Neoplasms – Immunology. 3. Immunity. QU58-133]
QR186.145 574.2′9 76-6133
ISBN-0-87819-077-5

This book represents information obtained from authentic and highly regarded sources. Reprinted material is quoted with permission, and sources are indicated. A wide variety of references iare listed. Every reasonable effort has been made to give reliable data and information, but the author and the publisher cannot assume responsibility for the validity of all materials or for the consequences of their use.

International Standard Book Number 0-87819-077-5

Library of Congress Card Number 76-6133
Printed in the United States

PREFACE

This is the second monograph in the Uniscience Series on Immunology and Lymphoid Cell Biology. A principal aim of this series is to highlight certain areas of research in immunology and related fields which appear to offer exciting potential for research and which are of interest to a broad group of basic and clinical scientists.

This volume emphasizes studies on the role of RNA in various forms of the immune response. These considerations are timely in view of the growing interest in immunotherapy, particularly the use of cell-free extracts, RNA, and/or "transfer factor" to correct immunodeficiencies. Apart from the important practical aspect of this work, the separation and isolation of molecules which control, modify, or augment the immune response will be of value in determining how these molecules operate in vivo to exert their immunologic effects. It is safe to assume that the application of modern techniques of molecular biology to the crude materials, presently being employed for immunotherapeutic purposes, will provide more powerful tools for an understanding of the molecular biology of the immune response and more potent therapeutic agents. Dr. Cohen's volume considers several aspects of this important problem and presents the current views on the state of the art in these areas.

A. Arthur Gottlieb
Editor-in-Chief

FOREWORD

It has been 20 years since reports on the enhanced immunogenicity of RNA extracts of antibody-forming tissues first appeared. The heightened immune response results, it is believed, from the association of retained antigens with RNA fragments to form a unique, cellular-dependent RNA-antigen complex.

One would expect that experimental findings of this sort, along with the extraordinary findings that in some instances the extracts possessed "informational" properties capable of phenotypically transforming responding cells to new functions, would be greeted by the community of immunologists with great hospitality. This has not been the case in every instance. In part, it may be the result of suspicions generated in the minds of scientists on hearing of experimental findings out of the "mainstream" of accepted concepts. One theoretical immunologist is reported to have commented as the observations were confirmed independently that the findings would have to be dealt with because "they won't go away." On the other hand, in part at least, reticence on the part of the scientific community toward accepting the role of biologically active RNA in the antibody response may result from two additional, significant problems – the data itself and the difficulty for many in repeating the observations. In some experiments the differences between the effect of biologically active RNA upon the cells of experimental and control groups, i.e., those exposed to RNA and those incubated without, have been small, if not marginal. In some instances less than twofold effects have been observed (see Cohen, Table 2). It may be the result of an inherent property of the system itself beyond the reach of the investigator. In others, independent confirmation of the results has been difficult even where experimental protocols have been followed most carefully. One persistent group accepted 2 years of failures before success was achieved. The "fragility" of the active material may be involved; the ubiquitous presence of ribonucleases is a continual obstacle. Even in the hands of the most successful experimentalists, the reaction succeeds inconstantly, reflecting unknown and uncontrolled variables in the described procedure. It is, no doubt, a most complicated cellular reaction.

At last, a scientific climate appears to be changing. The notion that immune RNA is a natural component of at least part of the antibody response and a therapeutic device of potentially great importance is gaining acceptance. The observations themselves are no longer called into question; what is debated is the precise mechanism involved, the characterization of the active substance(s), etc. This volume is presented in this spirit – in the hope that reviews of the reaction involving both clinical and basic immunology, as well as the nonspecific adjuvant properties of oligonucleotides and the remarkable effects of transfer factor, will further describe these incompletely understood phenomena, suggest new experimental approaches, and call to light the overall importance of the reaction.

Edward P. Cohen
Editor

THE EDITOR

Edward P. Cohen, M.D., is Associate Professor at La Rabida-University of Chicago Institute, the Departments of Medicine and Microbiology, University of Chicago and the Biological Sciences Collegiate Division, University of Chicago.

Dr. Cohen attended the University of Miami (Florida) and received his M.D. in 1957 from Washington University (St. Louis).

Dr. Cohen is a member of the American Association of Immunologists, American Society for Cell Biology, American Academy of Allergy, Academy of Medicine (New Jersey), American Society for Microbiology, Central Society for Clinical Research, Chicago Association of Immunologists, Chicago Society of Allergy, and the Reticuloendothelial Society. He is presently on the Committee on Immunology at the University of Chicago and is Public Affairs Representative for the American Association of Immunologists.

Dr. Cohen's current research interests include genetic controls governing the expression of cell membrane-associated antigens of malignant cells and the effect of specific antiserum on the expression of membrane-associated antigens of murine leukemia cells. He has published more than 90 research papers and abstracts.

CONTRIBUTORS

Peter Baram, Ph.D.
Senior Scientific Officer
Division of Immunology
Hektoen Institute for Medical Research
Professor
Department of Microbiology
University of Illinois at the Medical Center
School of Basic Medical Sciences
Chicago, Illinois

Peter J. Deckers, M.D.
Assistant Professor of Surgery
University Hospital
Boston University Medical Center
Boston, Massachusetts

Arthur G. Johnson, Ph.D.
Professor
Department of Microbiology
The University of Michigan
Ann Arbor, Michigan

David H. Kern, Ph.D.
Assistant Research Oncologist
University of California
Los Angeles, California

John A. Mannick, M.D.
Professor and Chairman
Division of Surgery
Boston University School of Medicine
Surgeon-in-chief
University Hospital
Boston, Massachusetts

Yosef H. Pilch, M.D.
Associate Professor of Surgery
University of California at Los Angeles
Chief, Surgical Oncology
Harbor General Hospital
Torrance, California

Bosco S. Wang, M.S.
Research Assistant
Department of Surgery
Boston University School of Medicine
Boston, Massachusetts

TABLE OF CONTENTS

"BIOLOGICALLY ACTIVE" RNA AND THE IMMUNE RESPONSE*

Edward P. Cohen

TABLE OF CONTENTS

INTRODUCTION

In 1957 Garvey and Campbell[1] discovered the presence of antigen fragments complexed with small moleucles of RNA in the urine of immunized rabbits. The complexes appeared after antigen injection and persisted in gradually decreasing amounts for prolonged periods (a matter of weeks or months) after immunization. Injected into normal, nonimmunized recipients, these RNA-antigen complexes possessed adjuvant properties, i.e., they stimulated a brisk and vigorous immune response in excess of what was observed upon the injection of similar or even larger quantities of antigen alone.

The biological significance of RNA-antigen complexes in the immune response and the explanation of their immunogenic properties are questions still not resolved. Although Garvey and Campbell's investigation was stimulated by the then popular notion that antibodies were formed around antigen fragments acting as templates or patterns to guide the folding of immunoglobulins to a complementary three-dimensional structure (according to this theory, the persistence of antigen, or antigen derivatives, would be required for continued antibody synthesis), their suggestion that RNA might somehow be a normal accompaniment of antibody formation initiated a series of investigations on the mechanism of antibody synthesis, as well as tumor immunology, organ transplantation, cell-mediated immunity, and cellular differentiation. While it is safe to say that most immunologists now accept the notion that RNA or RNA-antigen complexes transferred from one cell type to a second stimulate a specific response in the recipient cell, it is perhaps safer to say that the characterization of the active substance is incomplete, the mechanism is uncertain, and, more importantly, its precise role in the immune response is unclear.

In this chapter, I will review several basic lines of evidence. I shall assume that two RNA-associated cellular products are involved in the experimental observations thus far described. The first, along the lines of Garvey and Campbell's early observation, is that a fragment of the antigen originally injected retaining specific antigenic

*Supported by Grant No. GB-40659 from the National Science Foundation and the Leukemia Research Foundation, Inc.

determinants and complexed intracellularly by enzymatic processes is immunogenic in precursors of antibody-forming cells, stimulating specific antibody formation. In this scheme, specificity of the response resides in the associated fragments of antigen; the RNA plays a secondary role, acting perhaps as a "carrier." The second product is RNA alone, free of antigen. Here, the specificity of the response resides in the RNA itself, which may have informational "messenger" properties, entering responsive cells, combining with ribosomes present, and directing the synthesis of specific immunoglobulins.

A third product, resulting from the fortuitous, nonenzymic association of antigen or antigen fragment with RNA, shall be described as well. This complex forms, possibly extracellularly, as a result of nonspecific ionic interactions between RNA and the antigen. For unknown reasons, the complex is more immunogenic than the antigen alone – possibly the result of the heightened affinity of certain cell types for the RNA. In any event, according to this argument, RNA-antigen complexes play no biologically significant role in the normal course of the immune response. The heightened immunogenicity of the complex results from the fortuitous attraction of precursors of antibody-forming cells for antigen associated with RNA.

THE ROLE OF RNA-ANTIGEN COMPLEXES IN THE INDUCTION OF ANTIBODY SYNTHESIS

Crude cellular extracts of RNA (from a variety of cell types, including sources as diverse as bacteria and yeast, as well as tissues from various organs of immunocompetent animals) form "tight" complexes with carefully defined antigens. A vigorous and heightened specific immune response results after such "artificial" RNA-antigen complexes are injected into immunocompetent recipients.[2,3] "Synthetic" antigens, as enzymatically synthesized copolymers of tyrosine, glutamine, and alanine covalently associated with a poly-L-lysine "backbone" (TGA-L), provoke specific and clearly defined immune responses. Complexes of RNA and TGA-L, prepared simply by mixing the antigen with RNA, in the absence of cells, enzymes, or other cell products proved to be "stronger" antigens than the same antigen without RNA.[4] They provoked a heightened and more prolonged antibody response (Table 1).

The disturbing feature of these observations for

TABLE 1

Priming Activity in CSW and C3H Mice of RNA in Complex With or Mixed With TGAL

Priming injection	Mouse strain	% ^{125}I-TGAL bound by serum[a]				
		0–10	11–25	26–50	51–75	76–100
PE cell minor band RNA	CSW	14				
	C3H	6				
E. coli minor band RNA	CSW	9	1			
	C3H	6				
PE cell major band RNA (3 μg)	CSW	6	1	3		
+TGAL (1 μg)	C3H	6				
PE cell major band RNA (3 μg)	CSW	10	1			
	C3H	4				
E. coli major band RNA (3 μg)	CSW	2	4	3	6	3
+TGAL (1 μg)	C3H	6				
E. coli major band RNA (3 μg)	CSW	6				
+TGAL (0.1 μg)	C3H	6				
E. coli major band RNA (3 μg)	CSW	10	1			
	C3H	6				

[a]25 μl of 1:10 dilution of antiserum mixed with 0.0025 μg of TGAL.

From Goodman, J. W., Roelants, G. E., and Byers, V. S., *Ann. N.Y. Acad. Sci.*, 207, 288, 1973. With permission.

proponents of the hypothesis that RNA plays an essential role in the immune response is that the adjuvant properties were observed even if the RNA is obtained from nonmammalian sources. Clearly, it is difficult to accept the notion that RNA-antigen complexes are essential features of the normal course of immunity if bacterial RNA can substitute in the reaction for RNA from the lymphoid cells of immunocompetent animals.

Sedimenting in density gradients, these "artificial" RNA-antigen complexes formed bands at densities ranging from 1.56 to 1.63, lighter than RNA free of antigen banding at approximately 1.68. The complexes formed most effectively if the materials used as antigens, possessed a net positive charge,[5] emphasizing the nonspecific nature of the complex.

It is difficult to interpret in detail the mechanism of the adjuvant effect of RNA-antigen complexes formed in this manner; too little is known of the precise cellular mechanisms involved in the induction of the antibody response, i.e., the recognition of antigens as foreign substances. Even the molecular form of the antigen at the moment it induces immunoglobulin synthesis is unknown. It is likely a solubilized derivative of the material injected. The adjuvant properties of RNA-antigen "mixtures" might result from the enhanced attractiveness of antigens complexed to RNA for phagocytic cells or perhaps from the nonspecific stimulatory effects of RNA on the reticulo-endothelial system.[6] The injection of digests of RNA alone, without antigen, leads to a small but detectable increase in the antibodies formed to a variety of defined antigenic substances. Other "explanations" equally vague are plausible. In another chapter, Johnson presents a detailed review of the adjuvant properties of oligoribonucleotides administered to immunocompetent animals together with or separately from antigenic stimulation. The problem is that these carefully described nonspecific effects of polynucleotides on the antibody response severely complicate the interpretation of the work of numerous other investigators in the field. One cannot be certain whether reports of the biological activity of RNA extracts prepared from antibody-forming tissues, such as spleen and lymph nodes as well as the liver of immunized animals (also the source of biologically active RNA extracts and, while not a primary organ of the antibody response, possessing an extensive network of phagocytic cells), result from the presence of retained antigen fragments, along with the nonspecific adjuvant properties of RNA; whether the RNA, perhaps free of antigen contamination, plays a more essential role; or whether both phenomena are operative in any given experiment. Furthermore, cell-dependent complexes of RNA and antigen are described (*vide infra*), brought together by specific and enzymic processes in phagocytic cells even by investigators who later emphasize the nonspecific nature of the complexes.[7]

There is little doubt that antibody-forming tissues of immunized animals selectively and preferentially retained antigen for prolonged periods in immunogenic form.[1,8-10] ^{35}S-labeled bovine serum albumin, a soluble antigen, was detected in rabbit liver for 140 days after intravenous injection. Extracted from the liver along with RNA, it provoked a vigorous and specific immune response in nonimmunized rabbits.

Originally, Garvey and Campbell found RNA-antigen complexes in the urine of immunized animals. Their appearance in the urine indicates that they circulate in the blood of the immunized animal and are small enough to filter through cells in the glomeruli of the kidneys.[11] Soluble antigens, such as bovine serum albumin, injected intravenously into rabbits are identified 5 hr later in the blood. At that time, variable quantities are found intact, fragmented, or complexed with RNA or oligoribonucleotides.[11] The appearance of antigen-RNA complexes in the urine is enhanced by a second injection of the antigen. The reaction to antigen injection is specific, i.e., injection of a noncross-reacting antigen does not stimulate the appearance in the urine of the antigen injected previously.[12,13]

Neither the role nor the source of the RNA-antigen complexes described by Garvey and Campbell in the immune response has been determined. Whether they are specifically synthesized by cells involved in antibody formation, play an active essential role in immunoglobulin synthesis, or are nonspecific "by-products" of cells and antigen has not been resolved.

Fishman[14,15] investigated the possible essential role of RNA in immunity using an in vitro system. He prepared a suspension of peritoneal cells from rats and exposed them to T_2 phage. The use of bacteriophage as antigen provided an extraordinarily sensitive test for antibody. RNA extracts prepared from the antigen-exposed cells incubated with rat lymph node cell explants led to the appearance of antiphage neutralizing substances. The reaction was specific – RNA extracts from cells incubated with T_5 phage failed to lead to the production of substances which neutralized T_2 phage.

Several comments in regard to this early work may be offered which apply to many later studies as well. While the suspension of peritoneal exudate cells consisted predominantly of macrophages, it was, nevertheless, heterogeneous and included lymphocytes among other cell types. Thus, it would not be possible with certainty to identify the cell of origin of the biologically active RNA. Immunized rats possess antiphage-neutralizing substances in the sera. An increase in such substances over "background" was observed and had to be considered in all experiments. It is unlikely that the RNA in the extracts prepared by Fishman retained informational properties for antibody synthesis; the sucrose density gradient patterns of biologically active RNA revealed that it was mostly degraded. Finally, the population of *responding* cells consisting of lymph node explants contained all cellular elements known to be involved in antibody formation, including macrophages. Thus, it is quite possible that antigen fragments contaminating the RNA preparations were "recycled" through macrophages stimulating antibody-forming cells in the lymph node explants *de novo.* It should be immediately pointed out, however, that the finding of such fragments *a priori* in no way diminished the importance of these observations. In later studies many of these critical comments have been investigated.

In Fishman's study the addition of antigen alone to the lymph node explants failed to lead to the initiation of antibody synthesis. At that time, the primarty initiation of antibody synthesis in vitro had not been accomplished. Furthermore, treatment of the extracts with ribonuclease abolished the biological activity of the preparations, i.e., the appearance of antiphage-neutralizing substances, suggesting that RNA was an active participant in the reaction. The concentration of ribonuclease used was large by biochemical standards, consisting of an enzyme:substrate ratio of approximately 1:100.

In such impure extract (consisting of multiple species of partially degraded RNA along with variable quantities of protein and DNA), however, the identification of intact antigenic determinants in the active preparations has little meaning. It indicates that the preparations are impure. No conclusions regarding their possible participation in the biological activity of the preparation may be reached by these methods, and an additional, more indirect approach is required.

Cohen and Parks[16] and Cohen et al.[17] studied the biological activity of RNA extracts of antibody-forming tissues. They found that extracts prepared from the spleens of mice immunized with sheep red blood cells converted a small but significant proportion of cells from the spleens of nonimmunized mice to antibody-forming cells. Antibody-forming cells were detected by the plaque-assay method. The reaction was specific, i.e., extracts prepared from the spleens of mice immunized with noncross-reacting antigens or from mice not deliberately immunized were inactive. Adding sheep red blood cells to spleens from nonimmunized mice before the extraction of RNA failed to yield active preparations. A "background" of spontaneous antibody formation had to be dealt with.

Cohen's extracts, as well as Fishman's, were impure and contained, in addition to RNA, both DNA and protein. However, the likely essential participation (if not elucidated) of RNA in the response was indicated by the extraordinary sensitivity of the active extracts to ribonuclease. Enzyme:substrate ratios as high as 1:700,000 inactivated the extracts (approximately 0.001 μg ribonuclease destroyed the biological activity of 700 μg of extract dissolved in saline in less than 10 min at 20°). Extracts dissolved in medium containing divalent cations were partially resistant to ribonuclease. In later studies Cohen found[18] that divalent cations aid in stabilizing the secondary structure of the RNA (Figure 1); this is indicated by the heightened hyperchromicity of RNA dissolved in buffer containing calcium and magnesium relative to that observed for aliquots of the RNA solution dissolved in buffer without calcium and magnesium. This observation deserves special emphasis. Later sections of this book will present the capacity of RNA extracts injected into immunocompetent recipients, including humans, to affect the immune status of the recipient. Ribonuclease is present in blood and intercellular fluids and would be expected to quickly degrade such RNA. RNA dissolved in solutions containing salts in physiologic concentrations resisted degradation. Persistence of its biological activity

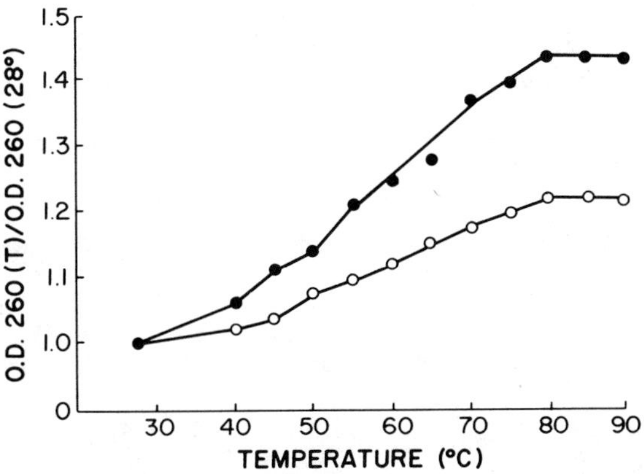

FIGURE 1. Hyperchromicity of RNA from the 8 to 12S fraction after heating. About 160 μg of biologically active RNA from the 8 to 12S fraction was dissolved either in medium 199 or in 0.15 *M* sodium chloride containing 2×10^{-4} *M* EDTA. Optical density at 260 nm was determined in a Zeiss spectrophotometer equipped with a flow-through heating device. (○) Calcium(II) and Magnesium (II); (●) + calcium(II) and magnesium(II). (From Cohen, E. P., *Nature,* 213, 462, 1967. With permission.)

was noted after incubation with ribonuclease if the reaction was performed in medium containing calcium and magnesium, but not if the reaction was performed in medium free of divalent cations.

RNA added to rabbit spleen cells is traced intracellularly. Based on the incorporation of isotopically labeled RNA, as many as 4×10^{10} molecules of rabbit RNA added to the medium were detected in the cytoplasm of individual cells, where it was resistant to degradation by ribonuclease persisting for several hours.[19] Data suggesting that rabbit cells might have a discriminatory capacity between mammalian and bacterial RNA were obtained by the observations that bacterial RNA was incorporated by rabbit cells to a lesser extent and was quickly degraded.

There is little doubt and abundant evidence that complexes of RNA and antigen are readily detectable in extracts of antibody-forming tissue. Injected into immunocompetent recipients, they provoke a vigorous and specific immune response. "Immunogenic" RNA has been found associated with T_2 phage, sheep red blood cell, and bacterial flagellar antigens, as well as chemically defined haptens. [13,20-22] The size of the associated RNA fragment in most studies is too small to allow it to have an "informational" role, i.e., acting as a messenger RNA for antibody synthesis.[6,15,23] It may originate in a macrophage, a cell believed to play an essential role in "preparing" antigen for inducing an antibody response in other cell types.[24]

It is likely that macrophages are at least one cellular source of complexes of RNA and antigen. In later studies Fishman[24] and others[25] found that extracts of purified or enriched cell suspensions of macrophages yield immunogenic RNA-antigen complexes.[24] Indeed, macrophages play a unique role in antigen processing. By electron-microscopic identification, approximately 10% of antigenic horseradish peroxidase ingested by rat macrophages is identified in "sequestered sites," other than lysosomal vacuoles, where it is retained and not degraded.[26] Antigen has been identified in subcellular "compartments"[27] or associated for prolonged periods with plasma membranes of the cells.[28,29] Antibody formation results upon the addition of immunocompetent, but not yet stimulated, cells to such macrophage suspensions.

Other classes of lymphoid cells yield immunogenic RNA. Cohen et al.[30,31] found that RNA extracts of the *thymus* stimulated a specific immune response upon addition to spleen cells from nonimmunized mice (Table 2). The thymus is not an antibody-*forming* organ, although it has a central role in the development of the immune capacity. Like other studies in this field, the precise characterization of the active substance in the extracts of thymic RNA was not determined. Whether antigen or antigenic derivatives are involved is unknown.

While it is clear that RNA extracts of immunologically reactive tissues have the capacity to stimulate specific immune responses in precursors of antibody-forming cells, conflicting evidence abounds as to whether the biologically active RNA-antigen complex is a naturally appearing cellular product of macrophages, released to specifically stimulate surrounding lymphoid cells to form antibodies as a natural accompaniment of the immune response, or an artifactual product, resulting from the spontaneous association of RNA and protein (antigen) fortuitously more immunogenic than antigen alone.

The finding of a unique, enzymically associated, cell-dependent complex of RNA and antigen would be necessary to support the concept that RNA plays a "natural" role in the course of the antibody response. Such a complex most appropriately should appear in phagocytic cells

TABLE 2

Biological Activity of RNA Extracts from the Thymus

Experiment[a]	RNA Source	RNA Quantity (μg)	Cells Source[b]	Cells Quantity ($\times 10^6$)	Background plaques	No. plaques > background
1	Thymus	500	p.e.	30	95	101
2	Thymus	40	Spleen	25	19	17
	Spleen	40	Spleen	25	19	16
3	Thymus	400	p.e.	63	18	70
	Spleen	2,300	p.e.	63	18	59
4	Thymus	83	p.e.	13	192	87
	Thymus after RNAse	83	p.e.	13	192	0
5	Thymus	700	p.e.	40	29	11
	Thymus after RNAse	700		40	29	0
6	Thymus	170	p.e.	15	12	13
	Thymus after RNAse	176		15	12	0
7	Thymus	1,200	Spleen	400	109	41
	Thymus after RNAse	1,200		400	109	0
8	Thymus	500	Spleen	40	31	35
		500	Thymus	40	2	0
	Spleen	2,000	Spleen	40	31	52
		2,000	Thymus	40	5	0
9	Thymus	1,200	Spleen	133	47	33
		1,200	Thymus	107	4	3
	Spleen	4,000	Spleen	133	48	77
		4,000	Thymus	107	4	4
	Liver	4,380	Spleen	133	45	0
10	Thymus	292	Spleen	352	5	10
	N-thymus[c]	1,000	Spleen	352	4	0
	Spleen	1,330	Spleen	352	5	14
11	Thymus	160	Spleen	85	28	0
		320	Spleen	85	28	29
		480	Spleen	85	28	30
	N-thymus[c]	158	Spleen	85	18	0
		317	Spleen	85	32	3
		475	Spleen	85	16	0
12	Thymus	15	Spleen	33	3	10
	Liver	15	Spleen	33	3	3
	Spleen	15	Spleen	33	3	10

[a]The results of 18 separate experiments are presented. Approximately 20 CD-1 mice were injected i.p. with SRBC 3 days before they were killed. RNA was extracted by the warm phenol method from either the spleen, thymus, or liver, as indicated. Cells were obtained from either the spleens or peritoneal cavities of CD-1 mice not injected with SRBC. After the cells were incubated with RNA for 5 min at 37°C, the proportion forming antibody to SRBC was determined by the plaque-assay method. In each experiment fractions were taken from a pool of cells. Variation between duplicate background determinations was less than 10%. In those experiments involving degradation of RNA with ribonuclease, digestion was carried out at 37°C for 30 min at an enzyme:substrate ratio of 1:100.
[b]p.e., peritoneal exudate.
[c]RNA from glands of normal mice.
[d]Immunized with RRBC.
[e]Indicator RBC.

From Cohen, E. P., Patt, D. J., Conrad, A. H., and Legler, M. L., *Ann. N.Y. Acad. Sci.*, 207, 83, 1973. With permission.

TABLE 2 (continued)

Biological Activity of RNA Extracts from the Thymus

Experiment[a]	RNA Source	RNA Quantity (μg)	Cells Source[b]	Cells Quantity (× 10⁶)	Background plaques	No. plaques > background
13	Thymus	117	p.e.	60	14	26
	Spleen	700	p.e.	60	13	40
	Liver	765	p.e.	60	14	3
14	Thymus	1,710	Spleen	313	33	17
	Thymus-RBC[d]	1,710	Spleen	313	26	0
15	Thymus	810	Spleen	225	31	16
	Thymus-RBC[d]	810	Spleen	225	24	0
16	Thymus	975	Spleen	100	7	7
		975	Spleen	100	0	0 (rabbit)[e]
17	Thymus	522	Spleen	100	20	17
	Thymus-RBC[d]	522	Spleen	100	15	0
		1,044	Spleen	100	20	5
18	Thymus	572	Spleen	110	42	45
		572	Spleen	110	0	0 (rabbit)[e]
	Thymus-RBC[d]	572	Spleen	110	27	0

and not in other cell types, e.g., nonlymphoid cells.

Gottlieb et al.[32-34] find a unique ribonucleoprotein product of rat macrophages sedimenting as a distinct band in cesium sulfate density gradients. Using the synthetic copolymer of glutamine, alanine, and tyrosine, they identified an antigen-RNA complex of approximately 12,000 wt, containing an antigen fragment of approximately 3,600 d "tightly" associated with RNA. The RNA itself had a high degree of secondary structure and was only slowly degraded by ribonuclease. A similar complex was not found in extracts of rat liver cells exposed to the same antigen.[35,36] Significant in determining whether the RNA-antigen complex is an enzyme-dependent cellular product or a nonspecific complex of the synthetic copolymer used as antigen is Gottlieb's finding that the L stereoisomer complexed to RNA more than eight times as efficiently as the D isomer. As the net charge of the D and L forms of the "synthetic" antigen is the same, the preferential association of the L form to macrophage RNA indicates an active cellular association occurs. It would be hard to understand how the basis for a preferential binding to the L form might occur if the association is nonspecific, possibly resulting from complementary charge differences between the RNA and antigen. Similar RNA-antigen complexes formed with crude cell-free homogenates of macrophages.[37,38] Others find RNA-antigen complexes appearing in rabbit liver cells after immunization and induced specific antibody synthesis in vitro in rabbit spleen cells.[39,40] The capacity of such complexes to induce antibody formation in unprimed spleen cells was abolished by ribonuclease, but not pronase.[41] However, while the enzyme sensitivity of the active extracts points toward an essential role for the RNA present, it shall be reiterated that antigen may still be required for the induction of antibody synthesis. The theoretical minimum number of antigen molecules required to induce antibody synthesis in an in vitro system may be quite small, perhaps as few as 10^7 to 10^{10} molecules.

The *insensitivity* of the active extracts to proteolytic enzymes, such as trypsin and pronase, has little theoretical bearing on the question of whether antigen is present. This experimental approach is unlikely to achieve convincing "negative" data.

A direct attempt was made by Schlager et al.[42] to detect antigen in biologically active extracts of RNA. They used an arsenic-containing organic compound as antigen and searched for the presence of arsenic in active extracts of RNA by atomic spectroscopy. To the limits of resolution of the method (5 pg in 500 μg of RNA), none was detected.

The sensitivity of the complex to ribonuclease

is of greater significance and points toward an essential, albeit undetermined, role for intact RNA. The induction and expression of an immune response could be accomplished in vitro in a brief period, using extracts of RNA from lymphoid cells exposed to antigen.[43]

RNA extracts from peritoneal cells or suspensions of spleen cells of normal nonimmunized mice exposed in vitro for only 30 min to heterologous red blood cells actively converted nonimmune mouse spleen cells to form antibody-forming cells. As previously, the proportion of responding cells was small, approximately 1 in 1 million. The antigen content of the active preparations was not determined; however, the activity of the extracts was abolished by prior treatment with ribonuclease. The entire reaction, from the addition of antigen to the appearance of antibody, occurred in less than 4 hr. It appeared to be an active induction event rather than release of preformed immunoglobulins, as the appearance of antibody-forming cells was inhibited by inhibitors of cellular metabolism. The addition to the cell suspension of antigen alone, under similar conditions, had no effect.

Other research groups confirm that RNA extracts from the spleens of immunized mice convert a small, but significant proportion of nonimmune spleen cells to antibody-forming cells and that the extracts, whose biological activity is abolished by exposure to ribonuclease, contain detectable quantities of antigen.[44,45] RNA from the spleens of mice immunized with bacteriophage induced the formation of specific antibody when added to spleen fragments of nonimmunized mice. Sedimentation of the RNA extracts through sucrose density gradients revealed the biologically active material sedimented in a range of 8 to 12S. Antigen fragments were identified in the active preparations, indicating that the preparations were not pure. Whether the antigen was essential for the success of the reaction could not be determined by this experimental approach.[46] Cohen et al.[17] found that the reaction was more successful if the RNA donor and the recipient cells were of the same inbred mouse strain. The significance here is that a strain-specific preference would not be expected if the RNA component of the active material acted only as a biologically insignificant carrier of antigen, its adjuvant effect perhaps resulting from an improved facility through which macrophages ingest and "process" antigen for the subsequent induction of the antibody response. Similar results are reported by others.[47]

The essential role of RNA in the reaction converting nonimmune spleen cells to cells forming antibody was strengthened by Smyth and Maurer's finding[48] that the association of "synthetic" polypeptide antigens with the RNA of rabbit peritoneal cells is a cell-dependent reaction, related to the net charge of the antigen, its state of aggregation, and (as Gottleib reported) its stereoisomeric form. Complexes preferentially formed intracellularly with L isomers rather than D. These data strongly support the concept that RNA-antigen complexes are involved in the induction of the antibody response. They indicate that intracellular mechanisms exist in lymphoid cells to form specific complexes. However, they do not say that the formation of RNA-antigen complexes by lymphoid cells is an obligate part of the immune response. One can easily imagine that multiple avenues to induction exist – RNA as an intercellular mediator is only one route toward that end. There is no reason to believe that antigen alone, not complexed to RNA or oligonucleotides, is unable to stimulate antibody formation directly; indeed, there is considerable evidence to indicate that *induction occurs without the intervention of RNA.*

Further evidence indicating that RNA is involved in the induction of the antibody response comes from experiments in nonresponder mice. Certain strains of mice "respond" to structurally defined synthetic copolymers of amino acids and other "defined" antigens forming specific antibodies, while other strains fail to respond to the same antigens. They do not respond to the same antigens by forming antibodies, although they are quite normal in every other respect. A sizable list of antigens which classifies mice, as well as other animal species, in this regard has been accumulated. Responsiveness, under genetic control, is inherited as a simple Mendelian trait. F_1 hybrids of "responder" and "nonresponder" strains "respond." As an example, C57B1/6 mice fail to form antibodies to the synthetic copolymer $Glu^{69} Ala^{36} Tyr^{16}$. This antigen is immunologically inert in such mice, insofar as antibody synthesis is concerned. Of significance, however, is the observation by Pinchuck et al.[49] that RNA extracts prepared from peritoneal cells (mostly macrophages) of responder strains exposed to the "synthetic" antigen $Glu^{69} Ala^{36} Tyr^{16}$ stimulated the appearance of specific antibodies in "nonresponder" mice.

Responsiveness appears to be dependent on the capacity of thymus-derived cells to "recognize" antigens as foreign, not on the presence or absence of structural genes in the genome of nonresponder strains for the synthesis of immunoglobulins. This conclusion is based on the successful induction of specific antibody synthesis in nonresponder animals to antigens, ordinarily not immunogenic in such animals, if they are complexed to immunogenic carriers, e.g., bovine serum albumin. The RNA extracts successful in inducing a specific antibody response in nonresponders contained approximately 0.2% antigen. The activity was abolished by ribonuclease.[48] RNA extracts from mice nonresponsive to $Glu^{69}Ala^{36}Tyr^{16}$ failed to stimulate antibody formation in nonresponders, although they did in responder strains. Thus, a *specific* intracellular event occurred, supporting the concept that these complexes were involved in the natural course of events leading to the formation of antibody, at least in part.

Extracts of RNA prepared from the spleens of immunologically tolerant animals converted nonimmune cells to antibody-forming cells. The "efficiency" of the extracts was equivalent to extracts prepared from nontolerant animals, capable of actively forming specific immunoglobulins.[50] RNA extracts from the spleens of mice tolerant to *Salmonella paradysenteriae* somatic antigen stimulated antibody formation in lymphoid cells of nonimmunized mice, even though the donor of the RNA was incapable of responding to such antigens.[51]

The relevance of these observations to the question of the biological significance of RNA-antigen complex formation is not entirely clear. It likely indicates simply that the "processing" of antigen in tolerant mice proceeds normally, even though the cells potentially capable of forming antibodies are suppressed. Phagocytic processes are grossly unimpaired in tolerant animals, although little evidence on the molecular basis of tolerance is available.

The question now is whether Roelants and Goodman's carefully determined "negative" findings, described previously, affect the importance of the findings of Gottlieb and others who describe a cellular-dependent RNA-antigen complex and its important role in the induction of the immune response. Evaluation of a "negative" experiment is difficult, no matter how carefully performed. Numerous investigators, not only those studying the role of RNA in the immune response, have described an essential role for macrophages in the early "processing" of antigen.[52-55] RNA-antigen complexes, whether or not macrophage derived, clearly are more immunogenic than similar quantities of antigen not associated with RNA. Problems in interpretation of the existing data arise when questions relating to the precise role of RNA in the immune response, not on its origin, arise.

AN INFORMATIONAL ROLE FOR IMMUNE RNA

The adjuvant properties of RNA extracts of antibody-forming tissues are abolished by ribonuclease, indicating that the RNA itself is an important component of the active material. None of the data presented to this point relates directly to the *function* of the RNA itself. Evidence has appeared indicating it has an "informational" role, coding for at least portions of the immunoglobulin molecule itself. This question was investigated using donor and recipient animals of different genotypes.

The most unequivocal approach toward an answer to the question of whether biologically active RNA codes for immunoglobulins is that taken by Adler et al.[56] These investigators, more than any others, initiated a series of investigations bearing most directly on the question of whether the RNA itself has informational properties. In their studies they extracted RNA from the peritoneal cells of rabbits homozygous at one allotype locus (Ab^4/Ab^4) and exposed them to lymph node cells of rabbits homozygous at a second allotype locus (Ab^5/Ab^5). The antibodies appearing had the characteristics of the RNA donor. (Table 3). The converse was true, e.g., RNA from Ab^5/Ab^5 rabbits led to the formation of immunoglobulins in lymph node cells of Ab^4/Ab^4 rabbits which were of the Ab^5 type. The immunoglobulins newly formed by the recipient's cells were characteristic of the RNA donor and were of the IgM class. IgG antibodies were formed as well, expressing the allotype markers of the recipient animal. This probably indicates that antigen frag-

TABLE 3

Allotypic Specificity of Antibody Against T_2-phage Produced in Culture

Experiment no.	Donor of: Peritoneal exudate cells	Donor of: Lymph node fragments	Incubation (days)	Neutralization of T_2-phage TC fluids preincubated with: Normal rabbit serum (%)	Anti-4 (%)	Anti-5 (%)
1	b^4b^4	b^5b^5	5	53	10	40
			10	38	34	0
2	b^4b^4	b^5b^5	5	34	3	44
			12	27	36	17
3	b^5b^5	b^4b^4	5	60	50	19
			13	38	0	35
4	b^5b^5	b^4b^4	4	44	39	0
5	b^4b^4	b^5b^5	5	34	4	30

From Adler, F. L., Fishman, M., and Dray, S., *J. Immunol.*, 97, 554, 1966. With permission.

ments or RNA-antigen complexes were present in the peritoneal cell extracts and stimulated antibody formation *de novo*.

There is essentially only one explanation for Adler's observation, namely, that RNA with messenger properties formed in peritoneal cells exposed to antigen entered the cytoplasm of the recipient's lymphoid cells and associated with the ribosomes present, directing the synthesis of new immunoglobulins. (I am excluding other possible interpretations for which there is no known biological parallel, such as the possibility that the RNA acts as an "inducer" of immunoglobulin formation is pseudohomozygotic rabbits.) Results analogous to those involving L-chain markers were found in rabbit for H-chain markers as well.[57] Peritoneal cells of rabbits homozygous at one H-chain allotype locus converted cells from rabbits homozygous at another H-chain locus to form immunoglobulins characteristic of the donor. As before, the active material in the extracts was sensitive to ribonuclease.[58,59]

The logistical aspects of the reaction are formidable. To explain these results, one must postulate that mRNA molecules for H chains and L chains enter the cell and direct the synthesis of their respective polypeptides. The RNA extracts, used in general, are not fractionated and consist of many different "species" of RNAs, only a small proportion of which are likely to be mRNAs for immunoglobulins. That the reaction succeeds at all may be taken as an indication that the recipient's lymphoid cells that respond in this manner are "programmed" in some way, possessing recognition sites for "external" RNA. In no instance have any of the investigators in this field reported the formation of a "hybrid" molecule, consisting of H chains of one allotypic specificity and L chains of another allotypic specificity, as might be expected to occur if both external and internal sources of RNA were present in the same cell. Furthermore, the quantities of immunoglobulins synthesized by the converted cells have been large enough to lead to precipitin reactions in agar gels. Relatively large quantities of the RNA of the donor must have formed in the responding cell.[59-61] In other systems[62] heterologous mammalian mRNAs introduced into undifferentiated cells led to expression of the differentiated function of the RNA donor, the effect persisting for long periods.

Using more sensitive methods to detect "surface," i.e., membrane-bound, immunoglobulins, RNA from mouse plasmacytomas was found to convert the usual membrane-bound immunoglobulins of lymphocytes to those characteristic of the donor.[63,64] The RNA molecules obtained from the plasmacytoma cell lines were obtained from the 12 to 23S fractions of sucrose density gradients. Actinomycin treatment of the responding cells had no effect on the capacity of RNA to convert their surface immunoglobulins, and, as before, the success of the reaction was abolished by prior treatment of the extracts with ribonuclease.[64] In some instances the effect of

RNA on the membrane-bound immunoglobulin receptors persisted for prolonged periods (a matter of weeks),[65,66] suggesting that the RNA might have replicative "virallike" properties.[67] The responding cells may have fortuitously captured messenger RNAs for immunoglobulins from the medium and utilized them in the synthesis of new immunoglobulins.

Schaefer, Fishman, and Adler selectively removed spleen cells reactive to the antigen T_2 phage by passing them through an antigen-containing column. The cells which failed to bind to the column, nonreactive to T_2 phage, also failed to respond to RNA-antigen complexes isolated from the lymphoid cells of immunized animals. They *did* form T_2 antibodies if exposed to fractions containing informational RNA, suggesting that a small proportion of cells in the suspension freed of antigen-reactive cells selectively captured sufficient informational RNA to form immunoglobulins. These data might be interpreted as supportive of the concept that the cells responded to informational RNA nonspecifically, i.e., as "innocent bystanders." However, further studies by the same investigative group indicate that this interpretation is unlikely. In subsequent studies, Alder and Fishman[69] report that cells from allotypically suppressed rabbits (forming immunoglobulins of only one or two codominant alleles) fail to respond to informational RNA coding for the suppressed allotype marker. These data might be interpreted as indicating that the overall reaction is far more complicated and significant than generally realized.

RNA EXTRACTS FROM LYMPHOID CELLS SENSITIZED FOR CELL-MEDIATED IMMUNITY CONVERT NONIMMUNE CELLS TO STATES OF SPECIFIC SENSITIZATION

In 1964 Mannick and Egdahl first reported[70] that RNA extracts prepared from lymph nodes of immunologically mature rabbits actively rejecting skin allografts converted lymphoid cells from nonsensitized rabbits to active cellular immune states. Nonsensitized rabbits injected with RNA-exposed cells rejected allografts in a manner similar to rabbits who had rejected similar allografts previously, i.e., according to second-set kinetics. The reaction was specific for tissues of the same histocompatibility type as that used to sensitize the donor of the RNA. The injection of cells exposed to RNA from rabbits rejecting tissues of one histocompatibility type had no effect on the capacity of the recipient to reject tissues of an unrelated histocompatibility type. The grafts were rejected, but the time required for rejection to occur was the same as that required by animals not injected with cells, i.e., according to first-set kinetics.

As in previously described analogous studies involving the effect of RNA extracts on antibody formation in lymphoid cells, the active material in Mannick and Egdahl's study was sensitive to ribonuclease; its possible antigen content was not determined. Injection of RNA extracts alone, without cells, into nonsensitized recipients was without effect. The recipients remained nonsensitized.

In subsequent studies Mannick and Egdahl's findings were confirmed independently, initially by Sabbadini and Sehon.[71,72] Not only was transplantation immunity from sensitized cells induced, but enhanced graft-versus-host reactivity could be induced under similar circumstances as well. Sabbadini and Sehon[73] injected parental lymphoid cells from inbred mice into F_1 recipients. The appearance of a graft-versus-host reaction, detected by a spleen-assay method in newborn F_1 mice, was significantly enhanced by prior exposure of the parental cells to RNA extracts from lymphoid cells obtained from animals actively rejecting allografts of the *same* genotype as the recipients. The activity of the extracts was abolished by ribonuclease.

There can be little doubt of the validity of these experimental observations, carefully designed and performed. In many instances the results were read "blind" to avoid the prejudicial interpretation of sometimes marginal effects in a complicated system.[74] The problem is that as yet so little is known of the precise nature of the active material, and its mode of action, that a detailed understanding of the mechanism is impossible.

Fortunately, the development of in vitro procedures for the detection of cells sensitized for cell-mediated immunity has provided additional means of studying the reaction. The most commonly used in vitro "correlate" of cell-mediated immunity is the effect of soluble factors released from sensitized cells exposed to specific antigens on the migration of lymphoid cells (MIF). Using this method to detect the presence of sensitized cells, Jureziz et al.[75] and Thor and Dray[76] reported that peritoneal cells of nonsensitized guinea pigs were converted by RNA extracts from sensitized cells to react as sensitized cells exposed to antigen, i.e., they released MIF upon exposure to antigens, used to sensitize the donor of the RNA. The reaction was specific. As an example, incubation of cells exposed to RNA extracts from PPD-sensitized cells failed to convert cells to sensitivity to noncross-reacting antigens.

Although the active material in the extracts was not identified, as in similar experiments, it was inactivated by ribonuclease. Centrifugation of the RNA extracts in sucrose density gradients revealed that the "biologically active" material sedimented in the 6 to 12S range, similar to that described previously for RNA extracts active in converting nonimmune cells to form specific immunoglobulins. It is unknown whether antigen is a necessary accompaniment of the active material and whether sensitization results through a more rapid induction process than occurs without RNA. Systems similar to those used by Bell and Dray,[77] involving the release by converted cells of immunoglobulins of known allotype, are unavailable in these studies. Fractionation of the active RNA from the numerous species of nonparticipating RNAs present in the extracts, has not been accomplished, nor has its mode of action been determined.

It may be of special significance that the conversion of cells to cell-mediated immune states by RNA has been observed across species barriers. RNA extracts of lymphoid cells of rhesus monkeys sensitized to one antigen (e.g., PPD or coccidioidin) convert nonsensitive peritoneal exudate cells of guinea pigs[78] or peripheral blood cells of nonsensitive humans[79-81] to reactivity in the MIF assay system. RNA from nonsensitive donors is inactive, and reaction is specific for the antigen to which the donor of the RNA was sensitive. In these experiments, the active material was identified in the region of sucrose density gradients containing 4 to 18S RNA.

Arguments similar to those presented previously in regard to immunoglobulin formation apply here as well. It is difficult to imagine that mRNAs from sensitized cells, coding for what surely is a complicated cellular reaction, survive digestion with ubiquitous tissue and blood ribonucleases to enter sufficient numbers of cells of the recipient to effectively phenotypically "transform" them to new functions. This is especially true when one considers that the reaction is operative across species barriers. RNAs would have to penetrate cell membranes in sufficient quantities to allow altered cell function to occur. Even though mRNAs from rats, as one example, direct protein synthesis in cell-free systems as diverse as *E. coli*[82] (indicating that bacterial ribosomes accept and utilize mammalian mRNAs), it is, nevertheless, easier to accept the notion that antigen derivatives persist in sufficient quantities in the RNA extracts to stimulate cells in the recipient directly. "Natural" amplification of the reaction might occur as a result, involving such known agents as transfer factor. This is not to argue that the observations relating to the biological reactivity of RNA extracts of sensitized cells are trivial, unrelated to events occurring during the natural course of immunization. That the reaction is successful at all might be taken as strong evidence that the contrary is quite likely to prove to be the case. The potential clinical importance of these observations is apparent, and it is discussed in detail in subsequent chapters of this book.

PERSPECTIVES

The development of methods for the isolation and characterization of RNA molecules of a single molecular species makes it possible to investigate in precise terms the nature of biologically active RNA and RNA-antigen complexes. If, as seems likely through indirect evidence described previously, informational RNA (e.g., messenger RNAs for immunoglobulins) is the active material, then it may be possible in the near future to prepare such antibody mRNA in virtually pure form, add it to cultures of lymphoid cells from nonimmunized recipients, and observe, by sensitive and specific methods, the appearance of newly synthesized immunoglobulins characteristic of the RNA donor. In such an idealized experimental approach a negative result would have little significance. Informational RNA may not possess many of the characteristics of messenger RNA, preventing its isolation by "conventional" means. New methods of isolation may have to be developed, reflecting unique aspects of its structure. "Naked" RNA added to cell cultures may be readily degraded.

The success of experimental protocols involving the addition of "informational" RNA to cell cultures may result from "protection" of the RNA through the presence of excessive amounts of nonantibody-specifying RNA, acting as a "carrier." On the other hand, if, as has been suggested, biologically active RNA is protected through its RNase-resistant secondary structure or through association with a "protective" virallike protein coat, it may remain intact, entering responsive cells whose own genomes are repressed insofar as antibody synthesis is concerned and directing the synthesis of newly formed immunoglobulins.

Biologically active RNA may have additional virallike characteristics. Persistence of the effects may be related to its self-replicative properties. Viral structures have been identified in antibody-forming cells.[83] RNA-dependent DNA polymerase enzymes have been associated with mouse lymphoid cells.[24] Whether or not viruses and unique enzymes are responsible for the long-lived effects of biologically active RNAs on lymphoid cells, exposed either in vivo or in vitro to such extracts, has not been established. It is likely, using established methods for the isolation of viral particles, that the question of whether viruses carry mRNA for immunoglobulins may soon be answered.

Of greatest significance for biologists studying the cellular and molecular mechanisms of the antibody response is whether biologically active RNA plays a "natural" role in the development of immunity, i.e., whether it is a concomitant of the sequence of events beginning with the introduction of antigens into immunocompetent hosts and ending with the appearance of specific antibody. One postulated role for such RNA would be to "amplify" the response through the recruitment of "uncommitted" cells receiving antibody mRNAs from cells responsive to antigenic stimulation. The "amplification" of cellular immunity through characterized agents, such as transfer factor, would be an obvious analogy, although in the case of antibody formation the requirement for the antigen specificity of the antibody molecules formed against such an undetermined but large number of different antigenic specificities would necessitate a more complicated intercellular mediator. The dangers of teleologic interpretations of experimental observations are all too apparent here. One investigative approach toward this question will be to determine the cell types responsive to biologically active RNA. If it is determined that only antibody-forming cells, i.e., B cells, respond to such RNA, then its role in the natural course of immunity would be strengthened, although not necessarily established. One could imagine that B cells, while receiving no "external" nucleic acid source of stimulation, fortuitously respond to antibody mRNA under experimental circumstances. If other lymphoid not ordinarily identified as antibody-forming cell types are found responsive, e.g., T cells and macrophages, then its natural role in the antibody response would be called into serious question, although its role as a potential therapeutic device, of course, would be no less significant. The successful treatment of cancer patients with materials not ordinarily involved in "natural" immunity is no less beneficial.

The capacity of RNAs with informational characteristics to modify the functions of a variety of cell types is not limited to viral- and immunoglobulin-specifying nucleic acids. Amos et al.[85] induced chicken fibroblast cells in culture exposed to heterologous and foreign RNAs to produce proteins antigenically cross reactive with those formed by *E. coli* or mouse myeloma cells. The response of the cells was specific for the foreign RNA present. Clearly, in this instance at least, the presence of "receptors" on responding cells for foreign RNA would be fortuitous, if they existed at all. It would not be, of course, a natural situation, reflecting instead the indiscriminate use by mammalian cells of informational RNA from diverse sources. As a further indication of the lack of discriminatory capacity of mammalian cells, for foreign RNA, *syngeneic* mouse skin grafts were rejected by inbred hosts after they were exposed in vitro to allogeneic RNA, indicating that, at least in part, alteration of the usual antigenic characteristics of the cells had occurred.[86-88]

There can be little doubt that this evidence, indicating that externally induced phenotypic alterations in the function and structure of cells may take place in experimental situations, will significantly complicate efforts to determine whether biologically active RNA plays a natural role in the development of immune competence.

The effects of RNA from one differentiated population of cells upon another undifferentiated population are not limited to the immune system.[89] Analogous effects have been observed involving RNA extracts prepared from hormone stimulated tissues, as one additional example.[90,91] There is no suggestion in this instance that RNA substitutes for hormones as an intercellular mediator of induced differentiation.

REFERENCES

1. **Garvey, J. S. and Campbell, D. H.,** The retention of S^{35}-labeled bovine serum albumin in normal and immunized rabbit liver tissue, *J. Exp. Med.,* 105, 361, 1957.
2. **Roelants, G. E., Goodman, J. W., and McDevitt, H. O.,** Binding of a polypeptide antigen to ribonucleic acid from macrophage, HeLa and *Escherichia coli* cells, *J. Immunol.,* 106, 1222, 1971.
3. **Roelants, G. E. and Goodman, J. W.,** The chemical nature of macrophage RNA-antigen complexes and their relevance to immune induction, *J. Exp. Med.,* 130, 557, 1969.
4. **Goodman, J. W., Roelants, G. E., and Byers, V. S.,** RNA:antigen complexes: Mechanism of formation and the testing of a postulated mode of action, *Ann. N.Y. Acad. Sci.,* 207, 288, 1973.
5. **Egan, M. L., Smyth, R. D., and Maurer, P. H.,** Association of polypeptides with ribonucleic acid, *J. Immunol.,* 107, 540, 1971.
6. **Johnson, H. G. and Johnson, A. G.,** Regulation of the immune system by synthetic polynucleotides, *J. Exp. Med.,* 133, 649, 1971.
7. **Roelants, G. E. and Goodman, J. W.,** Immunochemical studies on the poly-γ-D-glutamyl capsule of *Bacillus anthracis, Biochemistry,* 7, 1432, 1968.
8. **Saha, A., Garvey, J. S., and Campbell, D. H.,** The physiochemical characterization of the ribonucleic acid-antigen complex persisting in the liver of immunized rabbits, *Arch. Biochem. Biophys.,* 105, 179, 1964.
9. **Garvey, J. S., Campbell, D. H., and Das, M. L.,** Urinary excretion of foreign antigens and RNA following primary and secondary injection of antigens, *J. Exp. Med.,* 125, 111, 1967.
10. **Campbell, D. H. and Garvey, J.,** The nature of retained antigen and its role in immune mechanisms, *Immunology,* 3, 261, 1963.
11. **Yuan, L., Garvey, J. S., and Campbell, D. H.,** Molecular forms of antigen in the circulation following immunization, *Immunochemistry,* 7, 601, 1970.
12. **Yuan, L. and Campbell, D. H.,** *Immunochemistry,* 9, 1, 1972.
13. **Yuan, L. and Campbell, D. H.,** Antigen processing and formation of antigen-RNA conjugates, *Ann. N.Y. Acad. Sci.,* 207, 337, 1973.
14. **Fishman, M.,** Antibody formation in vitro, *J. Exp. Med.,* 114, 837, 1961.
15. **Fishman, M.,** Induction of antibodies in vitro, *Annu. Rev. Microbiol.,* 23, 199, 1969.
16. **Cohen, E. P. and Parks, J. J.,** Antibody production by nonimmune spleen cells incubated with RNA from immunized mice, *Science,* 144, 1012, 1964.
17. **Cohen, E. P., Newcomb, R. W., and Crosby, L. K.,** Conversion of nonimmune cells to antibody-forming cells by RNA:strain specificity of the response, *J. Immunol.,* 95, 583, 1965.
18. **Cohen, E. P.,** Conversion of nonimmune cells into antibody-forming cells by RNA, *Nature,* 213, 462, 1967.
19. **Wang, S. R., Giacomoni, D., and Dray, S.,** Physical and chemical characterization of RNA incorporated by rabbit spleen cells, *Exp. Cell Res.,* 78, 15, 1973.
20. **Archer, S. J. and Wust, C. J.,** Comparison of immunogenic RNA extracted from peritoneal exudate cells and Kupffer cells of the rat, *Ann. N.Y. Acad. Sci.,* 207, 241, 1973.
21. **Gottlieb, A. A.,** The antigen-RNA complex of macrophages, in *Nucleic Acids in Immunology,* Plescia, O. J. and Braun, W., Eds., Springer-Verlag, New York, 1968.
22. **Fishman, M., Adler, F. L., and Rice, S. G.,** Macrophage RNA in the in vitro immune response to phage, *Ann. N.Y. Acad. Sci.,* 207, 73, 1973.
23. **Fishman, M. and Adler, F. L.,** The role of macrophage-RNA in the immune response, *Cold Spring Harbor Symp. Quant. Biol.,* 32, 343, 1967.
24. **Fishman, M., Adler, F. L., and Rice, S. G.,** Macrophage RNA in the in vitro immune response to phage, *Ann. N.Y. Acad. Sci.,* 207, 73, 1973.
25. **Askonas, B. A. and Rhodes, J. M.,** Immunogenicity of antigen-containing ribonucleic acid preparations from macrophages, *Nature,* 205, 470, 1965.
26. **Catanzaro, P. J., Graham, R. C., and Schwartz, H. J.,** Ultrastructural identification of possible sites of antigen processing in macrophages, *J. Immunol.,* 103, 618, 1969.

27. **Kolsch, E. and Mitchison, N. A.,** The subcellular distribution of antigen in macrophages, *J. Exp. Med.,* 128, 1058, 1968.
28. **Unanue, E. R., Cerottini, J. C., and Bedford, M.,** The persistance of antigen on the surface of macrophages, *Nature,* 222, 1193, 1969.
29. **Unanue, E. R. and Cerottini, J. C.,** The immunogenicity of antigen bound to the plasma membranes of macrophages, *J. Exp. Med.,* 131, 711, 1970.
30. **Cohen, E. P., Majer, J., and Friedman, K.,** The response of the thymus in the adult mouse to immunization, *J. Exp. Med.,* 130, 1161, 1969.
31. **Cohen, E. P., Patt, D. J., Conrad, A. H., and Legler, M. K.,** The involvement of reiterative DNA in the antibody response, *Ann. N.Y. Acad. Sci.,* 207, 83, 1973.
32. **Gottlieb, A. A., Schwartz, R. H., Kudva, S. A., and Waldman, S. R.,** The nature of antigen-ribonucleoprotein complexes, *Ann. N.Y. Acad. Sci.,* 207, 279, 1973.
33. **Bishop, D. C. and Gottlieb, A. A.,** Distribution of antigen-ribonucleoprotein complexes within rat peritoneal exudate cells, *J. Immunol.,* 107, 269, 1971.
34. **Gottlieb, A. A. and Schwartz, R. H.,** Antigen-RNA interactions, *Cell. Immunol.,* 5, 341, 1972.
35. **Gottlieb, A. A.,** Studies on the binding of soluble antigens to a unique ribonucleoprotein fraction of macrophage cells, *Biochemistry,* 8, 2111, 1969.
36. **Gottlieb, A. A.,** The RNA-antigen complex of macrophages, in *Nucleic Acids in Immunology,* Plescia, O. J. and Braun, W., Eds., Springer-Verlag, New York, 1968, 471.
37. **Gottlieb, A. A. and Straus, D. S.,** Physical studies on the light density ribonucleoprotein of macrophage cells, *J. Biol. Chem.,* 224, 3324, 1969.
38. **Gottlieb, A. A.,** Macrophage ribonucleoprotein: nature of the antigenic fragment, *Science,* 165, 592, 1969.
39. **Duke, L. J. and Harshman, S.,** In vitro induction of anti DNP antibody by immunogenic RNA from rabbit liver, *Immunochemistry,* 8, 431, 1971.
40. **Harshman, S., Duke, L. J., and Six, H.,** Mechanism of antibody biosynthesis, *Immunochemistry,* 6, 175, 1969.
41. **Duke, L., Miller, C., and Harshman, S.,** The in vitro induction of antibody formation with immunogenic RNA, *Ann. N.Y. Acad. Sci.,* 207, 145, 1973.
42. **Schlager, S. I., Dray, S., and Paque, R. E.,** Atomic spectroscopic evidence of the absence of a low molecular weight (486) antigen in RNA extracts shown to transfer delayed-type hypersensitivity in vitro, *Cell, Immunol.,* 14, 104, 1974.
43. **Mosier, D. E. and Cohen, E. P.,** The induction and rapid expression of an immune response in vitro, *Nature,* 219, 969, 1968.
44. **Herscowitz, H. A. and Stelos, P.,** The induction of bacteriophage-neutralizing antibody by mouse spleen immunogenic RNA, *J. Immunol.,* 105, 779, 1970.
45. **Herscowitz, H. B. and Stelos, P.,** The induction of bacteriophage-neutralizing antibody by mouse spleen immunogenic RNA, *J. Immunol.,* 105, 779, 1970.
46. **Herscowitz, H. B. and Stelos, P.,** Immunogenicity of mouse spleen RNA, *Ann. N.Y. Acad. Sci.,* 207, 104, 1973.
47. **Juras, D. S. and Abramoff, P.,** In vitro induction of sulfanilic acid-specific antibodies by RNA extracted from rat spleen, *J. Immunol.,* 105, 1244, 1970.
48. **Smyth, R. D. and Maurer, P. H.,** In vitro formation of polypeptide-RNA complexes, *Ann. N.Y. Acad. Sci.,* 207, 301, 1973.
49. **Pinchuck, P., Fishman, M., Adler, F. L., and Maurer, P. H.,** Antibody formation: initiation in "nonresponder" mice by macrophage synthetic polypeptide RNA, *Science,* 160, 194, 1968.
50. **Friedman, J.,** Immunogenic RNA during immunity and tolerance, *Ann. N.Y. Acad. Sci.,* 207, 178, 1973.
51. **Friedman, H.,** The nature of immunogenic RNA-antigen complexes in immune and tolerant mice, in *Nucleic Acids in Immunology,* Plescia, O. P. and Braun, W., Eds., Springer-Verlag, New York, 1968, 505.
52. **Bishop, D. C., Pisciotta, A. V., and Abramoff, P.,** Synthesis of normal and "immunogenic RNA" in peritoneal macrophage cells, *J. Immunol.,* 99, 751, 1967.
53. **Friedman, H. P., Stavitsky, A. B., and Solomon, J. M.,** Induction in vitro of antibodies to phage T2: Antigens in the RNA extract employed, *Science,* 149, 1106, 1965.
54. **Friedman, H.,** Antibody plaque formation by normal mouse spleen cultures exposed in vitro to RNA from immune mice, *Science,* 146, 934, 1964.
55. **Wilson, D. B. and Wecker, E. E.,** Quantitative studies on the behavior of sensitized lymphoid cells in vitro, *J. Immunol.,* 97, 512, 1966.
56. **Adler, F. L., Fishman, M., and Dray, S.,** Antibody formation initiated in vitro, *J. Immunol.,* 97, 554, 1966.
57. **Bell, C. and Dray, S.,** Conversion of nonimmune rabbit spleen cells by ribonucleic acid of lymphoid cells from an immunized rabbit to produce IgM and IgG antibody of foreign heavy chain allotype, *J. Immunol.,* 107, 83, 1971.
58. **Bell, C. and Dray, S.,** Conversion of nonimmune spleen cells by ribonucleic acid of lymphoid cells from an immunized rabbit to produce γM antibody of foreign light chain allotype, *J. Immunol.,* 103, 1196, 1969.
59. **Bell, C. and Dray, S.,** Lymphoid cells converted by lymphoid RNA extracts in vitro and in vivo to synthesize allogeneic immunoglobulins, *Ann. N.Y. Acad. Sci.,* 207, 200, 1973.

60. **Bell, C. and Dray, S.,** Conversion of nonimmune rabbit spleen cells by ribonucleic acid of lymphoid cells from an immunized rabbit to produce IgG antibody of foreign light chain allotype, *J. Immunol.,* 105, 541, 1970.
61. **Bell, C. and Dray, S.,** RNA conversion of lymphoid cells to synthesize allogeneic immunoglobulins in vivo, *Cell. Immunol.,* 6, 375, 1973.
62. **Gurdon, J. B., Lingrel, J. B., and Marbaix, R.,** Message stability in injected frog oocytes: Long life of mammalian α and β globin messages, *J. Mol. Biol.,* 80, 539, 1973.
63. **Bhoopalam, N., Yakulis, V., Giacomoni, D., and Heller, P.,** personal communication.
64. **Giacomoni, D., Yakulis, V., Wang, S. R., Cooke, A., Dray, S., and Heller, P.,** In vitro conversion of normal mouse lymphocytes by plasmacytoma RNA to express idiotypic specificities on their surface characteristic of the plasmacytoma immunoglobulin, *Cell. Immunol.,* 11, 388, 1974.
65. **Yakulis, V., Bhoopalam, N., Giacomoni, D., and Heller, P.,** RNA and surface immunoglobulins of lymphocytes in plasmacytoma, *Fed. Proc.,* Abstr. 3152, 33, 1974.
66. **Boopalam, N., Yakulis, V., and Heller, P.,** Monoclonal IgM surface immunoglobulin on lymphocytes of aging NZB mice and its induction in young mice by RNA, *Clin Exp. Immunol.,* 16, 243, 1974.
67. **Katzmann, J., Giacomoni, D., Yakulis, V., and Heller, P.,** Characterization of two plasmacytoma fractions and their RNA capable of changing lymphocyte surface immunoglobulins (cell conversion), *Cell Immunol.,* in press.
68. **Schafer, A. E., Fishman, M., and Adler, F. L.,** Studies on antibody induction in vitro, *J. Immunol.,* 112, 1981, 1974.
69. **Adler, F. L. and Fishman, M.,** In vitro studies on information transfer in cells from allotype-suppressed rabbits, *J. Immunol.,* 115, 129, 1975.
70. **Mannick, J. and Egdahl, R. H.,** Transfer of heightened immunity to skin homografts by lymphoid RNA, *J. Clin. Invest.,* 43, 2166, 1964.
71. **Sabbadini, E. and Sehon, A. H.,** Transfer of transplantation immunity with RNA from lymphoid organs of immunized animals, in Proc. 1st Int. Congr. Transplant. Soc., Paris, 1967, 56.
72. **Sabbadini, E. and Sehon, A. H.,** Acceleration of allograft rejection induced by RNA from sensitized donors, *Int. Arch. Allergy Appl. Immunol.,* 32, 55, 1967.
73. **Sabbadini, E. and Sehon, A.,** Induction of accelerated graft rejection and enhanced graft-versus-host reaction with RNA from lymphoid organs of animals immunized with allogeneic tissues, in *Nucleic Acids in Immunology,* Plescia, O. and Braun, W., Eds., Springer-Verlag, New York, 1968, 560.
74. **Likhite, V., Sabbadini, E., and Sehon, A.,** Transfer of transplanation immunity with RNA, *Ann. N.Y. Acad. Sci.,* 207, 389, 1973.
75. **Jureziz, R. E., Thor, D. E., and Dray, S.,** Transfer with RNA extracts of the cell migration inhibition correlate of delayed hypersensitivity in the guinea pig, *J. Immunol.,* 101, 823, 1968.
76. **Thor, D. E. and Dray, S.,** Transfer of cell-mediated immunity by immune RNA assessed by migration-inhibition, *Ann. N.Y. Acad. Sci.,* 207, 355, 1973.
77. **Bell, C. and Dray, S.,** Synthesis and secretion of allogeneic immunoglobulins by lymphoid cells cultured in vitro after interaction with allogeneic RNA lymphoid cell extracts, *Fed. Proc.,* 32, 955, 1973.
78. **Paque, R. E. and Dray, S.,** Interspecies "transfer" of delayed hypersensitivity in vitro with RNA extracts, *Ann. N.Y. Acad. Sci.,* 207, 369, 1973.
79. **Paque, R. E. and Dray, S.,** Monkey to human transfer of delayed hypersensitivity in vitro with RNA extracts, *Cell. Immunol.,* 5, 30, 1972.
80. **Paque, R. and Dray, S.,** Transfer of delayed hypersensitivity to nonsensitive human leukocytes with Rhesus-monkey lymphoid RNA extracts, *Transplant. Proc.* 6, 203, 1974.
81. **Paque, R. E. and Dray, S.,** Interspecies "transfer" of delayed hypersensitivity in vitro with RNA extracts, *Ann. N.Y. Acad. Sci.,* 207, 369, 1973.
82. **Mach, B. and Vasselli, P.,** Template activity of RNA from antibody-producing tissues, *Science,* 150, 622, 1965.
83. **Giacomoni, D. and Katzmann, J.,** personal communication.
84. **Jackertz, D.,** Flow of information and gene activation during antibody synthesis, *Ann. N.Y. Acad. Sci.,* 207, 122, 1973.
85. **Amos, H., Askonas, B., and Soeira, R.,** Evidence for messenger of alien ribonucleic acid in chick cells, *Natl. Cancer Inst. Monogr.,* 13, 155, 1964.
86. **Guttman, R. D. Kraus, E. D., and Dolan, M. F.,** Immunology: rejection of isogeneic murine skin grafts, following exposure to allogeneic ribonucleic acid, *Nature,* 203, 196, 1964.
87. **Groth, C. F., Porter, K. A., Daloze, P. M., Huguet, C., Smith, G. V., Brettschneider, L., and Staezl, T. E.,** Effect of ribonucleic acid perfusion on canine kidney and liver homograft survival, *Surgery,* 64, 31, 1968.
88. **Lemperle, G.,** Prolonged survival of skin allografts after incubation with recipient DNA or RNA, *J. Surg. Res.,* 8, 511,
89. **Beers, R. F., Jr. and Tilghman, R. C., Eds.,** *Cellular Modification and Genetic Transformation by Exogenous Nucleic Acids,* Johns Hopkins University Press, Baltimore, 1973.
90. **Villee, D. B.,** Ribonucleic acid: Control of seroid synthesis in endocrine tissue, *Science,* 158, 652, 1967.
91. **Fujii, T. and Villee, C.,** Stimulatory effects of RNA on the growth of the seminal vesicle of immature rats, *Proc. Natl. Acad. Sci. U.S.A.,* 57, 1468, 1967.

THE ADJUVANT ACTION OF SYNTHETIC POLYNUCLEOTIDES ON THE IMMUNE RESPONSE

Arthur G. Johnson

TABLE OF CONTENTS

INTRODUCTION

Polynucleotides are produced by the action on the respective mononucleotides of an enzyme, polynucleotide phosphorylase, isolated from *Micrococcus lysodeikticus.* In the presence of this enzyme, the mononucleotide groups are joined via 3′-5-phosphodiester linkages and polymerized. When the polynucleotides of two base pairs are mixed (e.g., polyadenylic acid with polyuridylic acid, poly A:U, or polyinosinic acid with polycytidylic acid, poly I:C) annealing occurs, and a helical conformation ensues. Copolymers with varying proportions of the different nucleotides (e.g., poly A-U or poly I-C) also can be synthesized in this manner.

The availability of these synthetic polynucleotides offered a defined model for study of the mode of action of cellular nucleic acids. Their potential biological activity was explored beginning in 1966, when Dr. Werner Braun's group transferred to mammalian cells their interest and extensive knowledge of the biodynamic effects of bacterial DNA and DNA digests on division in bacteria;[1] at the same time our own group was following up on studies which showed a striking effect of syngeneic and allogeneic nucleic acids and their oligonucleotide breakdown products on antibody formation in inbred mice.[2,3]

Augmentation of the immune response by nucleic acids had become a focus for study earlier with the demonstration by Jaroslow and Taliaferro that nucleic acid-rich yeast extracts possessed the capacity to restore immunological responsiveness in X-irradiated rabbits.[4] Confirmation that nucleic acids and low molecular weight oligonucleotides were the active functional components was shown subsequently when preparations of both DNA and RNA, enzymatically degraded in vitro by specific nucleases, partially restored antibody synthesis to SRBC.[5] A similar restoration occurred in X-irradiated rats.[6] Purines, pyrimidines, nucleosides, and nucleotides were ineffective. In contrast to the elevation of titers seen in immunodeficient animals, an adjuvant activity of these preparations on the normal antibody response could not be detected in rats and rabbits in the above studies. However, using purified protein antigens in inbred mice as a model system, both DNA and RNA, as well as their respective oligonucleotides following nuclease actions, were shown to be fully capable of exerting an adjuvant action on normal immune responses.[2] Subsequently, when synthetic polynucleotides became available, it was demonstrated by Braun and Nakano that these complexes, likewise, could stimulate the normal immune response by increasing the number of antibody-forming cells some four- to five-fold.[7] Interest rapidly focused on these complexes because of their ready availability and more defined chemical constitution, and over the past 8 years they have been subjected to in-depth scrutiny. The early studies in this area have been reviewed;[8-19] consequently, this review will emphasize those studies published since 1970.

HUMORAL IMMUNITY

General Characteristics

The effect of polynucleotides on the immune response in vivo was characterized in detail by Schmidtke and Johnson.[20] The initial findings of Braun and Nakano[7] were confirmed using a different system – antibody formation to the purified protein antigen, bovine gamma globulin (BGG) – in the inbred Balb mouse. Poly A:U or poly I:C, in the 100- to 300-μg range, was shown to shorten the induction period leading to antibody appearance in the serum by 4 days and raise antibody levels some five- to tenfold. Adjuvant action was comparable to that previously shown for the potent Gram-negative endotoxins. None of the single-stranded polymers alone (poly A, poly U, poly I, or poly C) was active in vivo, indicating the complex form was required. The finding that antigen concentrations as low as 10 to 1 ng consistently induced significant antibody titers in the presence of poly A:U attested to the considerable potency of this adjuvant. The adjuvant signal in vivo probably was exerted very rapidly inasmuch as normal mouse serum inactivated poly A:U in as short a time as 10 min. This most likely could be attributed to the RNAse activity in normal mouse serum, since a 1:7.5 dilution of pancreatic RNAse sufficed to abrogate the adjuvant property also.

Poly A:U was found active in several animal species (i.e., mice, rats, guinea pigs, and rabbits), and its adjuvant activity was established for a wide variety of antigens,[20] notably bovine gamma globulin, bovine serum albumin, human gamma globulin, ferritin, sheep red blood cells, burro red blood cells, and isologous seminal antigens.[21] Both IgM and IgG were elevated, as well as an anaphylactic type of antibody in the mouse.[22]

In Vitro

The adjuvant action of polynucleotides in in vitro systems also has been established. The effect appears to be concentration dependent, with high doses inhibiting antibody-forming cells.[23] While Ishizuka et al. found poly A:U stimulated antibody formation in vitro at concentrations as low as 0.0001 to 0.001 μg and inhibited at concentrations of 0.1 μg,[24] others have reported stimulation over a dose range of 100 to 0.01 μg.[25,26] Different lots of the homoribopolymers may account for this discrepancy, as well as the use of spleen cells from different strains of mice. Although the complex form was required in the in vivo studies, a hint that the single-stranded poly A might be functional in vitro was seen in the studies of Ishizuka et al.[24] This was attributed to the absence of enzymes in vitro which rapidly degrade single-stranded polynucleotides in vivo. Exposure of the mouse spleen cells to poly A:U for the first

24 hr in culture was required for antibody enhancement, thus establishing the effect in vitro also as being exerted on an early event in the sequence leading to the immune response.[27]

Theophylline, an inhibitor of phosphodiesterase activity which thereby stabilizes intracellular cAMP levels, has been shown to potentiate the immunoenhancing effects of poly A:U.[28] The effect of theophylline also was concentration dependent, active doses lying in the 0.001- to 0.01-μg/ml range. In the main the inclusion of theophylline permitted the use of a 10- to 100-fold reduction in the concentration of poly A:U needed for maximal effects.[29] These findings raised the question as to whether the membrane target of poly A:U is the enzyme adenyl cyclase. The latter has been established in recent years to convert ATP into cyclic AMP, which in turn activates a variety of kinases involved in cellular synthesis. By prolonging the time for breakdown of cyclic AMP through inhibition of phosphodiesterase activity, an enhanced signal could be delivered.[17] Similar to the in vivo studies, poly A:U was capable of generating an immune response in vitro to very low doses of antigen, i.e., as few as 300 SRBC, whereas at least 300,000 were required in control cultures without-poly A:U.[30]

It is of interest that in a single investigation of the fate of antigen under the influence of poly A:U, a poly A:U dose-dependent decrease in uptake of the antigen, ferritin, by mouse thoracic duct lymphocytes was seen in vitro.[14]

Toxicity

Significantly, although both poly A:U and poly I:C are in the same class of complexes, they exhibit decided differences in toxicity. The toxicity of poly I:C has been well documented,[31-33] while poly A:U in the limited tests run to date has displayed no toxicity.[14] Both complexes, however, when given with adjuvant are antigenic in mice. Poly I:C given intravenously is lethal for mice and rabbits at LD_{50} levels of approximately 15 to 20 mg/kg and 0.5 to 1 mg/kg, respectively, and is pyrogenic for rabbits, dogs, and humans. Marked histopathologic changes in the surface epithelium of the small intestine have been observed, as well as atrophy of the lymphoid tissues of mice.[32]

A preliminary comparative study of the toxicity of poly I:C, poly A:U, and endotoxin again has verified the toxicity of poly I:C and the absence of such effects in poly A:U preparations.[14] Treatment of mice with actinomycin D reduced substantially the LD_{50} dose of endotoxin. In such actinomycin D-treated mice, the LD_{50}'s for poly A:U, poly I:C, and endotoxin were 50, 1, and 0.1 μg, respectively. Poly I:C was pyrogenic in rabbits, whereas poly A:U was only marginally so. In addition, poly I:C both prepared for and provoked the local Shwartzman reaction, whereas poly A:U was negative in this respect. Inasmuch as the adjuvant action of poly A:U was comparable to that of poly I:C and endotoxin, without the toxicity of the latter two, poly A:U has received increased attention recently.

Physicochemical Requirements

Little is known of the physicochemical requirements for activity of the complexes. It has been shown that neither repeated freeze thawing nor a 2-week incubation at room temperature had an adverse effect on the adjuvant action; however, on heating to 100°C this property was lost.[20] In addition, decreasing the hydrogen bonding of the respective complex constituents reduced its adjuvant action. Neither purine nor pyrimidine mono- or oligonucleotides could replace their respective polymers in the complex.[20]

Thermal activation, however, has been found to increase markedly the ability of the alternating copolymer, poly A-U, to induce resistance to vesicular stomatitis virus in human skin fibroblasts.[34] Thus, incubation of poly A-U for 5 min at 37°C changed the minimal concentration required to reduce by 50% virus plaque formation to 0.000004 μg from greater than 1 μg when incubated at 0°C. The inactive random copolymer (A,U) was not affected by this treatment and remained inactive, probably because a rigid helical structure was not possible without proper base pairing.

The molecular weight of the polynucleotides might be an important variable in the magnitude of the effects exerted by the homoribopolymers. It has been reported that the acute toxicity for mice of poly I:C decreased with diminishing molecular weight.[35] This was paralleled by its antivesicular stomatitis virus effect, which fell off drastically with a reduction in molecular weight below 100,000. Whether or not thermal activation or lowered molecular weight affects the adjuvant action of polynucleotides on antibody synthesis has not been reported as yet.

The fate of poly A:U added to immunocompetent cells is under investigation at present. Most of the previously acquired data on the fate of poly A:U involved study of nonimmunocompetent cells. In the latter, a difference between the two complexes, poly A:U and poly I:C, has been reported. Poly A:U entered Ehrlich ascites tumor cells without separation of base pairing and without breakage of strands, whereas poly I:C did not enter the tumor cells as an entity. Poly C remained at the surface, while a portion of the poly I entered the cell. Poly U alone only was adsorbed to cell walls, but did not enter.[36] Likewise, Kelly and Levy were unable to detect intact complexed poly I:C in intracellular RNA following its extraction and testing for the capacity to increase resistance to Sindbis virus in rat embryo cells.[37] The uptake of either A:U or U alone increased when the tumor cells were pretreated with DNAse or exposed to mixtures of DNAse and the polyribonucleotides.[38] In addition, both adsorption and entry increased with declining pH values, these effects being heightened at pH 5 as compared to pH 7.5.[39] The authors attributed the adsorption to a charge effect, inasmuch as pretreatment of the cells with poly L-lysine caused an increase in adsorption and entry of poly A:U and adsorption of poly U. Polynucleotide-polylysine complexes also enhanced uptake. Despite this enhancing influence, poly U still did not enter the cell, but poly A:U did so faster.[40] Ancillary evidence for this hypothesis was gained by Mayhew and Juliano, who demonstrated an increase in binding and uptake of high molecular weight homologous RNA to Ehrlich ascites carcinoma cells in the presence of poly L-lysine.[41] Thus, the change in charge on the cell surface following adsorption of the cation increased the uptake of molecules of opposite charge.

Along this same line, Colby et al. have reported that DEAE-dextran added with poly I:C to cultures of chick embryo cells stimulated remarkably the synthesis of interferon;[42] 10 μg/ml of the cation permitted detection of antiviral activity by 0.001 μg/ml poly I:C. This polycation increased the rate of uptake of P^{32}-labeled poly I:C into chick embryo cells by 20-fold and the extent of uptake of the polynucleotides by 25-fold. Although poly I:C and poly A:U were taken up at the same rate, the poor interferon inducer, poly A:U, was broken down more rapidly than poly I:C by intracellular nucleases. In their studies they were unable to enhance the antiviral activity of helical polynucleotides by thermal activation at 37°C, as discussed previously, with the exception of the alternating polynucleotide, poly I-C, where a 1- log reduction in virus yield was reported.

Immunosuppression

A striking effect of the time of administration of the polynucleotides in relation to injection of antigen was noted.[20] Thus, the adjuvant activity of the polynucleotides was restricted to a time period either several hours before or several hours after exposure to antigen. Suppression of antibody synthesis occurred when the polynucleotides were given 12 to 24 hr before antigen, a finding also observed with other adjuvants. This effect could be exerted by both poly A:U[20] and poly I:C.[43] The mechanism of this inhibitory effect is unknown. One clue was offered in the work of Stout et al., in which it was determined that antigen-binding cells first increased and then decreased in number when spleen cells were removed from primed mice and incubated with poly A:U.[44,45] The loss in antigen-binding cells could be attributed to the release of Ig receptors from the surface of the B cells on prolonged incubation with poly A:U. However, regeneration of surface proteins, at least in vitro, occurs within a 6- to 12-hr period, and whether or not early injection of poly A:U causes a loss in receptor sites at the time of antigen administration remains to be determined. A second possibility now under investigation in the author's laboratory is whether poly A:U enhances the activity of suppressor cells, such that the latter become more efficient over the 24-hr period following stimulus by poly A:U.

Large doses of poly A:U also have been reported to induce depression of the immune response in vitro, in contrast to the enhancement routinely seen with smaller amounts.[23] In this in vitro system, 100 ng/ml of poly A:U depressed the response by 50%, whereas 0.1 and 1 ng/ml were stimulatory. Theophylline enhanced both the stimulatory and depressive phases of the response.

Conversely, the adjuvant activity of poly I:C has been shown to be abrogated by a prior injection of bacterial endotoxin.[46] Thus, endotoxin given to mice 1 day before 200 μg poly I:C depressed the increase in plaque-forming cells normally caused by the polynucleotide. On the other hand, although poly I:C and endotoxin share several stimulatory and suppressive properties, three injections of 50 μg poly I:C on alternate days followed 48 hr later by endotoxin failed to induce tolerance to the lethal aspects of the latter.[47] Yet, normal mice pretreated with poly I:C exhibited complete hyporeactivity to interferon induction by endotoxin, but infection of mice with 0.2 ml of a 14-day culture of BCG (Phipps strain) potentiated the release of interferon by poly I:C and did not enhance sensitization to the lethal aspects of the polynucleotide.[48]

Prevention of immunosuppression (tolerance) of poly A:U has been reported by Kong and her co-workers.[49,50] Injection of poly A:U either simultaneously with or 3 to 12 hr after the tolerogen (centrifuged BGG) prevented the establishment of tolerance in Balb/c mice. Inasmuch as the adjuvant did not function when given 1 day after this T-dependent antigen (a time when only the T cell is tolerant), the effect was postulated to be exerted on the thymus cell. Later cell-transfer studies established more directly an effect on the cortisone-resistant T cells. Transfer of cortisone-resistant thymus or bone marrow cells from donor mice given tolerogen with or without poly A:U resulted in tolerance in those mice given thymus and bone marrow cells from animals without poly A:U and immunity in mice receiving T cells from poly A:U-injected donors.[51] Reversal of an already tolerant state by poly A:U has not been reported as yet.

In addition, depression of the antinuclear antibody response to the polynucleotides per se has been reported. Injection of poly I:C and cyclophosphamide into NZB/NZW mice with active lupus reduced serum antibody to RNA and DNA, indicating that immunological tolerance to nucleic acids via the polynucleotides can be established.[52] Others, however, were unable to induce tolerance in 2-month-old, ANA-positive NZB mice when 160 μg poly I:C was followed 1 day later by 75 mg cytoxan, repeated three times in 56 days.[53]

CELLULAR IMMUNITY

Polynucleotides also activated the cell-mediated limb of the immune response. A single injection of 200 μg poly I:C reduced the time for rejection of isografts between female and male C57B1/6 mice from 23 days in the control to 10 days in the adjuvant-stimulated mice.[54] Graft rejection was not hastened further by increasing the number of injections. In addition, poly A:U was shown to have the capacity to restore the rejection process in neonatally thymectomized Balb/aj mice following grafting with skin from C57B1/6j donor. Three injections of 600 μg poly A:U i.p. on days 1/2, 1, and 3 after grafting decreased the time to rejection from 24 days in nontreated mice to 13.5 days in poly A:U-treated mice.[55]

Stimulation of human peripheral blood leucocytes also was found to be a property of poly A:U.[56,57] Suboptimal concentrations of PPD, tetanus toxoid, and allogeneic cells added to previously sensitized human leucocytes were amplified markedly by inclusion of poly A:U in the culture medium, as measured by tritiated thymidine incorporation. In contrast, the addition of 50 to 100 μg poly A:U to cultures of PHA-stimulated leucocytes harvested after 24, 48, or 72 hr in culture depressed DNA synthesis.

Poly I:C has been shown to increase the graft-versus-host activity of spleen cells when 1.0 mg was given i.p. 3 days prior to their transfer into 1- to 5-day-old F_1 hybrid litters.[58] It was postulated that enhancement of this aspect of the immune response may account for the antitumor effects of poly I:C.

Further examples of the enhancement of cellular immunity by synthetic polynucleotides are included under the subject headings of Infectious Diseases, Immunologic Diseases, and Tumor Immunity (discussed below).

CELLS AFFECTED

The actual cell types affected by the synthetic polynucleotides appear to be multiple. Enhancing effects have been exerted on the stem cell, macrophage, and T and B cells.[12] It may be that such substances affect many cells, both immune and nonimmune, at the membrane level, signaling the cell to synthesize and secrete those substances with which it had been previously programed genetically to respond. In this vein, it has already been determined that nonimmunocompetent cells are affected in a positive fashion by poly A:U.[59]

Stem Cells

A slight increase in hematopoietic colony-forming cells in normal mouse bone marrow cultured in agar was evoked by 100 ng poly I:C or poly A:U in the presence of a colony-stimulating factor.[60] This was confirmed in vivo with the finding of an eightfold increase in colony-forming cells in the spleens of mice injected intraperitoneally with 100 ng poly I:C.[61]

In addition, a role for the stem cell in polynucleotide action has been unmasked recently with a different approach. In seeking to understand the role of the thymus in immunologic reactions, Sheid et al. have demonstrated the capacity of a variety of nonthymic agents, including poly A:U, to elicit the expression of the T-cell antigens (TL or Thy-1) on precursor cells in the bone marrow.[62] Such cells also acquired helper-cell activity in vitro. The authors postulated that T-cell differentiation could be induced not only by passage of stem cells through the thymus, but also nonspecifically by direct exposure to substances such as poly A:U. Whether or not contact at this cellular level is responsible for the adjuvant action of polynucleotides, particularly in immunodeficient animals, is an important question.

Macrophages

Several aspects of the total immunologic role of the macrophage have been enhanced by poly A:U. For example, in vitro addition of antigen to normal macrophages and subsequent transfer of the washed cells into syngeneic mice after a 90-min incubation results in substantial enhancement of the immune response by some as yet unknown mechanism. Addition of poly A:U to the antigen-macrophage mixture enhances this response still further.[63]

In addition, a cytotoxic effect of macrophages on tumor cells in vitro occurs following activation by a variety of toxic agents such as endotoxin and poly I:C.[64a] The nontoxic poly A:U also was found capable of activating macrophages to this role using a similar in vitro system of tumor immunity, but involving C58 mouse leukemia cells as target and syngeneic peritoneal exudate cells (PEC) as effector cells.[64b] Normal PEC harvested 4 days after injection i.p. of thioglycollate were ineffective at inducing release of ^{51}Cr from labeled leukemic cells. Incubation of 2×10^6 PEC for 18 to 24 hr with either lipopolysaccharide or poly A:U activated these cells to lyse target cells when added in a 10:1 macrophage/leukemic cell ratio. Optimal activity was observed with 50 μg poly A:U (specific cytotoxicity, 63%).* LPS detoxified by deacylation with 0.1 or 0.01 *M* potassium methylate activated PEC to 35 and 25% specific cytotoxicity, respectively, as compared to 40% for the parent nondetoxified LPS. In electron-microscopic study of the normal thioglycollate-stimulated and activated macrophages, the latter appeared larger, with many more lysosomal granules, and had an increased phagocytic capacity for staphylococci. Inhibition with selective antimetabolites suggested that RNA and protein synthesis were required for activation of macrophages by lipopolysaccharide, and it is conceivable that these would be necessitated for activation by poly A:U also.[64b]

T Cells

A major effect of the polynucleotides on the T cell has been reported. The clearest manifestation appeared in circumstances where T cells were diminished in numbers through thymectomy or antithymocyte globulin.[55] Thus, injection of poly A:U with antigen into 1-month-old mice thymectomized at birth restored antibody snythesis to normalcy. Its mode of action was hypothesized as being through amplification of residual numbers of T cells peripheralized prior to thymectomy. Indirect evidence in support of this was gained by experiments in which division of cells important to antibody synthesis in mice was found to be initiated 6 hr earlier under poly A:U influence.[65] In these mice, vinblastine (a drug which inhibits cell division) reduced antibody synthesis by 90% some 6 hr before mice receiving only the antigen. Direct evidence that poly A:U was acting on T cells was achieved through use of the irradiated mouse model, into which both T and B cells must be adoptively transferred in order to mount an immune response. In experiments in which either T or B cells were selectively incubated in vitro, with and without poly A:U prior to transfer, a tenfold increase in rosette-forming cells (rfc) was apparent only in the animals receiving T cells incubated with poly A:U. Furthermore, by reducing the numbers of T cells to 40,000 (an amount 2 logs below that minimal threshold necessary to initiate antibody synthesis), a response could be elicited on injection with poly A:U, once again suggesting amplification. B cells,

*The relationship of the observation to the effects described in detail by Pilch (page 71) has not been established as yet. Pilch's observations involving immune RNA are found to be specific for the class of tumor used as immunogen.

selectively exposed to poly A:U in like manner, were not increased in activity.[55]

Using nu/nu spleen cells as a source of B cells and nonadherent cells, 10 μg poly A:U again exhibited the remarkable property of activating as few as 10,000 thymus cells to initiate an immune response in such cultures, whereas 10^6 cells were required for a like response in nonpoly A:U-treated control cultures.[30] Antigen-binding cells of both T and B cell derivation, however, were amplified by 500 μg poly A:U in a parallel study using SRBC-immunized mice. A two- to threefold increase was seen in B-cell rosettes, while there was a tenfold increase in those of T-cell origin.[66]

Hamaoka and Katz in a precisely designed study have confirmed and extended the effect of poly A:U on T cells.[67] Using adoptive transfer techniques and the hapten-carrier system, DNP-KLH (in which the carrier effect is exerted on the T cell, while hapten affects the B cell), a marked adjuvant action by poly A:U on anti-DNP antibody was seen. Although both the homologous carrier and hapten ordinarily are required for secondary stimulus, poly A:U-exposed T cells effectively stimulated a secondary response when DNP-KLH primed cells were exposed to DNP attached to a different carrier (bovine gamma globulin). When the primed spleen cells were depleted of T cells by treatment with antitheta serum and complement prior to adoptive transfer, poly A:U failed to effect a response. Thus, the effect of poly A:U appears to be established on the T cell in this system also. In a similar study, essentially the same conclusion was reached by Yamashita and Kitagawa.[68]

Since the above evidence suggested that poly A:U functioned by increasing the numbers of T cells, thus amplifying helper activity, the mitogenic effect of poly A:U for mouse thymocytes was tested. However, the resulting data supported the conclusion that poly A:U was *not* mitogenic for thymocytes per se and only a weak mitogen for peripheral T cells in the spleen.[16,18] For example, single-cell suspensions of thymus tissue at a number of concentrations did not significantly increase their uptake of tritiated thymidine when doses of poly A:U from 10^{-4} to 20 μg were added without antigen to separate cultures at either 0, 3, 6, 12, 18, 24, or 48 hr. Slight, but consistent, stimulation was achieved by 5×10^{-2} to 5×10^{-3} μg poly A:U, but only when SRBC was added as antigen (125 to 130% of controls receiving SRBC alone). Ruhl et al. also detected only a very low mitogenic activity of poly A:U on mouse thymocytes.[69]

Inasmuch as thymocytes per se appeared unable to respond to poly A:U under those culture conditions, the mitogenic effect of poly A:U on the division of peripheral T lymphocytes in the spleen also was tested. In order to deplete the number of B cells in the spleen such that any effects might be attributable to T cells, 2.5 mg corticosteroid were injected i.p. into mice 2 days before removal of the spleen. 5×10^6 such spleen cells were cultured for varying periods of time with or without SRBC and/or 1 μg poly A:U/dish.[18] Beginning at 12 hr of culture, there was a variable, statistically insignificant increase in thymidine incorporation in tissues containing poly A:U, both in the presence and absence of SRBC. By 24 hr the cells given antigen alone showed a 38% increase in thymidine incorporation, whereas poly A:U alone produced a 24% increase. Under the combined stimulation of poly A:U and antigen, incorporation increased to 150% of the control value. Similar results were obtained when the same experiments were performed without the addition of calf serum. For example, thymidine incorporation by spleen cells in the presence of poly A:U with or without SRBC was 226 and 180% of the control, respectively. A significant point was that poly A:U alone, in the absence of serum or antigen, was weakly mitogenic to peripheral spleen cells.

An in vivo approach also was used, in which this adjuvant was given i.v. to Balb mice on day 0 with or without a concomitant injection of SRBC.[18] At 2 hr prior to sacrifice, thymidine was injected. Although no significant effect was observed 18 hr after poly A:U administration, a 260% stimulation of spleen cells was observed 2 days after poly A:U was given. Similar values were obtained in mice receiving poly A:U plus SRBC, and the stimulatory effects were still observable on days 3 and 4 following injection. No effect was measurable on thymus cells. Reduction of cortisone-sensitive lymphocytes by i.p. injection of this drug did not reduce or modify the mitogenic effects of poly A:U, indicating an effect on peripheral T cells. However, since poly A:U has been shown to be mitogenic in B cell-rich spleens, an effect restricted to T cells must still be accepted with caution.[69-71]

Since poly A:U did not appear to affect

profoundly the *numbers* of T cells, an effect of this polynucleotide complex on the *function* of individual T cells was sought. It had been shown previously that T cells stimulated by allogeneic cells, or by antigen following previous sensitization, released into the culture fluids a soluble factor(s) capable of increasing antibody-forming cells.[72] To determine whether poly A:U was capable of releasing such stimulatory factors in vitro, normal (C57B1 × C3H)F_1 thymocytes (10^7) were incubated with and without varying doses of poly A:U for 10 to 90 min, washed extensively, and cultured for 6 or 24 hr, respectively.[73] The harvested supernatant fluids were filtered and assayed by addition to 10^7 normal, syngeneic, spleen cells at a 25% final concentration with BRBC or SRBC as antigen on day 0. Direct and indirect PFC's were assayed 4 days later. With regard to IgM PFC, when the data from 17 experiments were normalized such that spleen cells receiving BRBC alone were assigned the value of 100, those cultures which received supernatant fluids from control thymocytes cultured for 6 hr without poly A:U had a value of 113 ± 43. Spleen cells receiving supernatant fluids from thymocytes incubated with poly A:U for 10 min (AUSF) had a value of 305 ± 107. Background PFC was at a level of 14 ± 10. Thus, from these data it was clear that poly A:U possessed the capacity for increasing the functional activity of the individual T cells.

Exposure of thymocytes to poly A:U was found to cause the release of AUSF after only 6 hr in culture. Inasmuch as normal, nonstimulated thymocytes also elaborated enhancing soluble factors, but only after 18 to 24 hr in culture, it was concluded that poly A:U's action was to shorten the induction period necessary for soluble factor production. Supernatant fluids harvested at 0 time of culture contained no enhancing activity despite thymocyte exposure to poly A:U. This finding argued against the possibility that poly A:U was being transferred in the supernatant fluids to the assay culture.

Since one of the major functions of the T cell in humoral immunity is to regulate the synthesis of IgG and since the T cell is a major target of poly A:U's adjuvant action, it was important to test the effect of AUSF on IgG synthesis. Accordingly, AUSF was harvested from 6-hr cultures of (C57B1 × C3H)F_1 thymocytes (10^7) that were incubated with poly A:U for 30 min and washed extensively. Following sterile filtration, AUSF at a 25% concentration was added with antigen to primary syngeneic spleen cell cultures on day 0. Indirect and direct PFC's were assayed on day 4.

It was found that cultures receiving no BRBC, poly A:U, or supernatant fluids had background levels of 7 ± 13 IgG PFC/10^6; those receiving only BRBC had 7 ± 10, indicating antigen could not stimulate IgG production in primary culture; those receiving BRBC + 1 μg poly A:U and no supernatant fluid had no IgG plaques, indicating that A:U added directly to cultures also could not stimulate IgG PFC in vitro. Spleen cells stimulated by BRBC and control supernatant fluids from thymocytes incubated 6 hr without poly A:U also showed no detectable IgG plaques, but the value for cultures stimulated by BRBC and AUSF was 118 ± 39 PFC/10^6 viable cells. Thus, AUSF enhanced IgG PFC in primary spleen cell cultures.

In other systems the induction or enhancement of PFC responses by T-cell factors is dependent on the time of factor addition relative to antigen and culture initiation. When AUSF was assayed for PFC enhancement by addition on day 0 or day 2, no essential difference was found in the degree of enhancement of either IgM or IgG PFC. The fact that late addition of AUSF enhanced both IgM and IgG PFC suggested that AUSF may not necessarily play a role only in the initiation process.

In order to postulate an in vivo role for AUSF, it was important to determine whether poly A:U could stimulate peripheral lymphoid tissue as well as thymus cells to produce soluble enhancing factors. Accordingly, spleen tissue was incubated with poly A:U along with thymocytes as a positive reference control. AUSF prepared from the spleen was found to selectively stimulate both IgG and IgM PFC. AUSF from thymus tissue was positive as expected.[73]

In other experiments AUSF was found to enhance the secondary response, and AUSF prepared from cortisone-treated mouse thymus was equally as effective as that from normal mice in stimulation of primary IgM and IgG. In addition, AUSF failed to reconstitute antitheta + C′-treated spleen cells or spleen cells cultured from nu/nu mice. Thus, the presence of T cells appeared to be a requirement for the enhancing action of AUSF. Neither mild heat treatment (56°C for 30 min) nor dialysis for 36 hr[73] had any effect on AUSF enhancing activity.

As a consequence of the data described above,

it is postulated that the increased antibody synthesis induced by the polynucleotide adjuvants results, in the main, from stimulation of T-cell secretion of endogenous enhancing factors. This hypothesis is rendered more plausible by the finding of Neiderhuber et al. that i.v. injection of either poly A:U or theophylline to mice results in the appearance in the blood stream within 2 hr of substances capable of stimulating strongly PFC in vitro.[74]

B Cells

In contrast to T cells, low numbers of B cells exposed to poly A:U in vitro did not prove capable of inducing antibody synthesis in X-irradiated mice receiving T cells in excess.[55] Similarly, when spleen cells of nu/nu mice lacking functional T cells were exposed to poly A:U in vitro, no response occurred.[30] However, effects of poly A:U logically interpreted to be on B cells have been published. For example, hemolytic foci of antibody-forming cells occur in the spleens of X-irradiated mice given spleen cells from donor mice injected with SRBC, with or without poly A:U. The number of such foci are thought to be related to the number of T cells in the donor preparation. Inasmuch as Campbell and Kind saw no increase in numbers of such foci, but did obtain an increase in B cells secreting antibody, an effect of poly A:U on B cells was concluded.[75] This is in keeping with the evidence discussed above where no division of thymus cells per se was induced by poly A:U. On the other hand, the increase of B cells reported by Campbell and Kind could have occurred as the result of an increase in a secreted product of T cells rather than by their division, and their data are compatible with the evidence of Bick and Johnson that poly A:U functions by inducing a secretory product.[73]

The B cell also was implicated as the target cell for poly A:U action in a study documenting the augmentation by poly A:U of antibody responses to a synthetic polypeptide in low responder strains of mice in which the deficiency is controlled by the IR-3 gene and not the IR-1 gene.[76] In this investigation the functional expression of graded numbers of bone marrow cells removed from donors injected with antigen plus poly A:U and transferred to X-irradiated recipients with excess T cells was approximately twice that of mice receiving bone marrow cells from mice unexposed to poly A:U. No effect on T cells in a like experiment was seen. These data might be subject to the same interpretation introduced above, wherein increased B-cell function could be induced by an effect of poly A:U on T cells secretion of a stimulatory product, rather than an increase in actual T-cell numbers.

Evidence for a mitogenic effect of polynucleotides on B cells has been obtained in a series of papers on polyanion stimulation by Diamantstein et al.[77-81] and by Scher et al.[70,71] Whether or not the mitogenic action on B cells is responsible for the adjuvant action of poly A:U, however, is obscure. A partial correlation has been made. In these studies, poly A:U, I:C, U, C, A, and I all were shown to stimulate tritiated thymidine uptake in spleen cells from a variety of strains of mice, with poly A and poly C effects being only marginal. Although poly U alone and poly I alone were mitogenic, they did not stimulate PFC's.

In an additional study, Ruhl et al. have shown that poly I:C and poly A:U are mitogenic for spleen cells of nude mice.[69] In contrast to B cells, thymocytes gave only a minimal response. The relationship of the mitogenicity and adjuvant action could not be determined from these data inasmuch as PFC was not determined.

In the study of Scher et al.,[70,71] poly I, poly C, and poly I-C were shown to be mitogenic, with poly I the most active. Anti-B cell serum (anti-MBLA) suppressed mitogenicity, whereas antitheta serum increased mitogenicity by approximately 50%. Nu/nu mice responded well. Thus, T cells were not required for mitogenicity, nor were they stimulated into tritiated thymidine uptake. Mitogenicity was found to be mouse-strain dependent, with C3H/HeJ and CBA/J strains exhibiting the greatest activity. In a later study, poly A, poly U, poly A:U, poly G, and poly X were found to increase mitosis in B cells.[71]

Poly I:C was found to be mitogenic also for human spleen cell cultures, as well as for certain strains of mice. Human peripheral blood cells were not affected. CD-1 and C3H strains were found superior to Balb/c strains in this respect.[82]

INFECTIOUS DISEASES

As one might expect, the polynucleotide complexes have proven to be effective at increasing nonspecific resistance to a variety of infectious agents. Their action at releasing interferon in vitro and in vivo, as well as their activation of the various components of the immune response, led to testing the capacity of these adjuvants to enhance resistance to a variety of viral agents. With the far greater efficacy of poly I:C over poly A:U at inducing interferon, the majority of these tests have been performed with the former. An excellent effect was noted against infection with pneumonia virus of mice (PVM) following intravenous administration of poly I:C, and a very good effect against Columbia SK virus and vaccinia virus was noted when the adjuvant was given both intranasally and intraperitoneally. In addition, protection was afforded against parainfluenzae 1 (Sendai) virus.[83]

The efficacy of poly I:C and poly A:U against influenza virus was increased markedly by incorporation into a water-in-peanut-oil emulsion of the aqueous vaccine.[84] The increased antibody levels persisted in guinea pigs for approximately 6 months and in monkeys for at least 11 months. The potentiation exhibited exceeded the additive effects of the polynucleotides and water-in-oil adjuvant tested alone.

A striking protective effect against rabies challenge following intravenous injection of poly I:C was seen in rabbits. One milligram of this polynucleotide complex given from 1 day before to 6 hr after challenge sufficed. Protection was not evident when given 1 day after challenge.[85] A similar protective effect was seen in mice as an adjunct to vaccination with inactivated rabies vaccine.[86]

With regard to bacterial infections, poly I:C in microgram amounts has been shown in mice to enhance the number of survivors following challenge with a variety of Gram-negative and Gram-positive organisms.[87] In the case of *Brucella*, 300 μg poly A:U rendered 3×10^8 bacterial cells in saline as immunogenic as 3×10^{11} cells in oil adjuvant against challenge with *Brucella abortus.*[88] The number of colonies per gram spleen was counted 42 days after vaccination (11 days after challenge). No antibody was found in protected animals; consequently, activation of cell-mediated immunity by poly A:U was postulated.[88]

Cheers and Cone found 250 μg poly A:U given i.p. to CBA mice at $\pm$ 1 hr in relation to 4×10^5 *Brucella abortus* (strain 19 cells) lowered the numbers of organisms in the spleen on day 10 by 3 $\log_{10}$. Injection of poly A:U 26 days after the organism did not affect the chronic course of the disease.[89]

Against intracellular-dwelling *Listeria monocytogenes*, the subcutaneous injection of either poly I:C or poly A:U 1 to 3 days before injection reduced profoundly the number of organisms in the liver and spleen and produced a 100% survival against this lethal challenge.[90] Since resistance against this organism is an attribute of an increased bactericidal capacity of mouse macrophages, it is possible the latter cell was affected. However, as has been shown by Mackaness, lymphocyte-secreted products activate such macrophages, and the polynucleotides also might have been acting through the former cell.[91] Ancillary evidence for a nonspecific effect of poly I:C exerted through macrophages was gained in a study by Remington and Merigan, who demonstrated protection in mice against *Listeria* organisms inoculated 3, 8, and 14 days after injection of poly I:C. Other polyanions also were effective.[92] On the other hand, the synthetic polynucleotides failed to increase the already potent action of mycobacterial RNA preparations against another intracellular dwelling organism, homologous challenge with the H_{37} R_v strain of *Mycobacteria tuberculosis.*[93]

Polymorphonuclear cells from peritoneal exudates of guinea pigs also have been shown to be affected in a positive manner by poly A:U. Thus, injection of poly A:U or endotoxin restored phagocytosis-associated defects, as well as partial bactericidal activity against *Escherichia coli* in PMN's following damage by X-irradiation.[94] The release of endogenous substances, such as endogenous pyrogen, from rabbit blood leukocytes also have been demonstrated following a 9-hr incubation of such cells with 30 μg poly I:C/ml.[95]

Significant protection of mice by poly A:U or poly I:C against *Toxoplasma gondii* required mixing with Freund's adjuvant. Poly C alone showed some protection.[96]

IMMUNOLOGIC DISEASES

The adjuvant properties of the nontoxic complex, poly A:U, might be utilized beneficially in human beings, not only as an aid to bacterial, viral, and tumor vaccines, but in attempts to restore immunocompetency in patients with immunological deficiency diseases. However, such application first demands awareness of any potential inducement or enhancing activity of poly A:U in the pathogenesis of immunological diseases, e.g., antigen-antibody complex glomerulonephritis in lupus erythematosus. One important question lies in the antigenicity of the polynucleotides per se and whether or not the humoral or cell-mediated immunity engendered against the synthetic complexes interacts with native nucleic acids. An excellent review on this subject has been published recently by Lacour et al.[97] Suffice it to say that antibodies to both poly A:U and poly I:C with or without carrier have been elicited in several species under a variety of conditions. Although cross reactions occur, different structural determinant groups may be defined, inasmuch as antibodies specific for a given DNA or RNA have been determined.[97] It is important to note, however, that the route of injection is important in that rabbits failed to elicit antibody when injected with poly I:C by either the oral or nasal route.[98] Both poly A:U and poly I:C were shown to be antigenic in complete Freund's adjuvant.[99] High levels of antibody were produced also when the complexes were given to guinea pigs in combination with methylated bovine serum albumin as carrier.

If heterologous antithymocyte serum was given with the polynucleotide antigen, the response was augmented, indicating this antigen was thymus independent.[100] Yet, when the polynucleotides were given in Freund's adjuvant, a prolonged IgG response was noted.[101] The latter response is thought to be largely under thymus control. These data raised the question as to whether antinucleic acid antibody was regulated by T suppressor cells and whether the loss of such suppressor cells occurs in NZB/NZW mice, which give rise to anti-poly I:C antibody spontaneously by the age of 5 months.[100]

In studying the immune response to polynucleotides of these New Zealand mice, which are highly susceptible to spontaneous autoimmune disease, antibodies to poly I:C were found to be readily induced without carrier or adjuvant, as compared to other normal strains of mice.[102,103] Antibodies to poly I:C in human beings with systemic lupus erythematosus (SLE), as well as several other diseases, were noted also. In addition, a precipitation reaction has been noted between poly I-C, poly I, and poly C and both fresh and stored human serum from patients with SLE.[104] C′1q was identified as one factor precipitating with poly I. Poly I had been noted previously to inactivate both C′1 and C′4.[105] On the other hand, specific immunologic tolerance to the synthetic poly I:C and poly A:U could be induced with the aid of cyclosphosphamide,[99,102] and a relative tolerance also could be induced by two or more injections of aqueous poly I:C in Balb/c mice.[101] This approach may be advantageous as a therapeutic procedure for the antinucleic acid antibody in systemic lupus erythematosus.

The enhancement by polynucleotides of autoimmune diseases under certain conditions has been established with the study of Gumbiner et al.[106] Typical lesions of experimental allergic encephalomyelitis were incited in guinea pigs when 200 μg poly A:U was combined with either whole spinal cord homogenate or myelin basic protein and the mixture incorporated into Freund's incomplete adjuvant. The data suggested that poly A:U potentiated the T-cell response in cell-mediated immunity to this antigen. In a later study, Paterson and Drobish confirmed that 1 mg poly A:U given in incomplete Freund's adjuvant with guinea pig brain stem plus spinal cord produced the clinical signs and lesions of EAE in 4/4 guinea pigs.[107] Poly A alone was inactive at 300 and 600 μg; however, uncomplexed poly U alone at the 600 μg level under the same conditions elicited lesions in 13/15 animals and clinical signs of the disease in 2/15. Poly C alone at the same dosage caused lesions in 2/3 animals. Thus, when given in Freund's incomplete adjuvant, the single-stranded pyrimidine polynucleotides alone appeared to function.

Supportive evidence that poly A:U can enhance delayed hypersensitivity was published by Casavant and Youmans.[108] It was shown that 200 to 600 μg poly A:U, again in Freund's incomplete adjuvant, permitted the induction of cell-mediated immunity to PPD in guinea pigs. Dermal reactions and histologic changes were measured, and the reactions could be passively transferred with spleen cells. As in the induction of EAE, poly U alone was capable of sensitizing animals to PPD, when mixed with Freund's adjuvant. In an additional study, it was established that delayed hypersenitivity could be induced in guinea pigs to the polynucleotides per se.[109] Sensitization with poly A:U in Freund's incomplete adjuvant resulted in production of migration inhibitory factor (MIF) when either poly A:U or poly U alone, but not poly A alone, was added to the sensitized peritoneal exudate cells. MIF also was secreted by poly A:U-sensitized PEC on the addition of mycobacterial RNA or whole H37Rv organisms. Sensitization with poly U alone produced PEC which secreted MIF in the presence of A:U, but not when poly A or poly U alone was added.

The immunodeficiency state existing in newborn or very young mice has been overcome by poly A:U.[110] Thus, 10^9 SRBC given with either 30 μg poly I:C or poly A:U to newborn C57B1/6j mice permitted antibody synthesis to occur in this previously anergic population. The deficiency in this instance may lie again in T cells, inasmuch as injection of syngeneic T cells alone or poly A:U alone into 10-day-old Balb mice also incited premature antibody formation to SRBC.[111] T cells plus poly A:U had a synergistic effect. It is postulated that poly A:U hastens the maturation of T cells when given to less than 2-week-old mice, perhaps through the release of a substance like AUSF (discussed previously)[73] or in a manner suggested by the studies of Scheid et al., where stem cells in the bone marrow acquired T-cell characteristics under the influence of poly A:U.[62] Preliminary testing of the first hypothesis has revealed that AUSF shares the capabilities of poly A:U in this respect.

A similar immunodeficiency may prevail in aged mice, inasmuch as a fivefold reduction in PFC was seen in 1-year-old Balb mice as compared to 3-month-old mice.[111] Injection of 5×10^6 thymocytes from 4-week-old mice i.p. with SRBC restored to normalcy both the IgM and IgG PFC levels of the aged mice. As with the very young mice, poly A:U alone, without thymocytes, also was found capable of restoration.

TUMOR IMMUNITY

Increasing host resistance against tumors via the immune system is an important thrust of present day cancer research. Poly A:U and poly I:C have been found to be effective adjuncts in this respect, especially in instances where the tumor load has been decreased by surgery or immunosuppressive drugs. The rationale for use of the polynucleotides is twofold, namely, their capacity to enhance both limbs of the immune response in an effort to increase immunity to the weakly antigenic tumors and, second, the ability of poly I:C to induce interferon, which might be operative against virus-induced tumors. As with other pharmacologic agents, tumors have been found to vary widely in susceptibility to the polynucleotides.

Levy et al. have summarized early animal studies on the antitumor action of poly I:C,[112] as have Webb et al.[113] Treatment with 150 μg poly I:C/mouse every other day lessened the size of an implanted reticular cell sarcoma on day 28 from an average of 6 mm^2 to near 0 and doubled to tripled the survival time with respect to other tumors.[112] Interferon (10,000 units daily) also reduced tumor size and prolonged survival time, but less so than poly I:C. Although interferon was a moderately good inhibitor of Moloney sarcoma virus in vitro, unlike poly I:C it had very little effect on survival time in vivo.

A weak effect of poly A:U and poly I:C (given in 300-μg amounts on day −1, 0, and +1) on growth of Rauscher leukemia virus tumors in Balb/c mice was found by Webb et al.[113] Theophylline potentiated the poly A:U effect slightly. Similarly, BDF_1 mice injected with 10^6 L 1210 ascites leukemia cells followed by 5 mg poly I:C/day, given i.p. or s.c. on days 1 to 7 had an increased life span of some 25 to 40%.[114]

Negative results were reported by Larson et al. with respect to the capacity of single or multiple doses of poly I:C to affect the growth of an SV_{40} tumor transplant or viral oncogenesis in hamsters.[115] In an earlier study poly I:C was found to reduce the occurrence of internal, but not subcutaneous, tumors occurring after injection of adenovirus 12 subcutaneously into newborn hamsters, but only if treatment was started prior to virus injection.[116]

Poly I:C was found by Fisher et al. to temporarily inhibit the growth rate of visible transplantation tumors (methylcholanthrene-induced fibrosarcoma and murine polyoma) in mice.[117] Immunosuppression of the host with X-irradiation or antilymphocyte serum did not eliminate the inhibitory effect of the polynucleotides on tumor growth, suggesting that the polynucleotides may activate small numbers of residual T cells. Alternatively, poly I:C might be operating in this instance through an entirely different mechanism. High doses of poly I:C were found in this study to induce lymphopenia and death. Similar results have been published by Kreider and Benjamin, testing the effects of poly I:C on growth of a rat mammary adenocarcinoma in normal and immunosuppressed rats and on growth of a B16 melanoma in normal and immunosuppressed mice.[118] Inasmuch as poly I:C inhibited growth of both tumors, but did not restore the capacity of ALS-treated hosts to reject skin allografts, it was concluded that the stimulation of T cells by poly I:C played little or no role in its inhibitory action.

A significant increase in mouse antileukemia virus antibody was reported (by Tennant et al.) to occur when a synthetic polynucleotide derivative, poly (2′-*o*-methyl-adenylic acid), was given to newborn Balb mice and followed 4 hr later by infection with a Moloney sarcoma virus-murine leukemia virus complex.[119] This derivative may have higher resistance to nucleases in vivo due to methylation. In an interesting approach, Fourcade et al.[120] have shown that DBA/2J mice immunized with poly I:C complexed with methylated BSA in complete Freund's adjuvant exhibited protection against challenge with Friend leukemia virus. In two experiments an average drop of 1.5 $\log_{10}$ in the 50% lethal dose of virus was measured in mice with anti-poly I:C antibodies.

In an in vitro study, it was shown by Wagner and Cone, that poly A:U enhanced the stimulation of mouse T cells into cytotoxic lymphocytes by cell-bound alloantigens.[121] When CBA thymus cells were either cocultured with Balb/c spleen cells and poly A:U or prestimulated with poly A:U prior to addition to the allogeneic spleen cells, development of cytotoxicity towards the DBA/2-derived Cr^{51}-labeled mastocytoma was accelerated, and the amount of antigen required to achieve maximal stimulation was reduced substantially. Doses of poly A:U in the 1- to 10-ng range were effective, while a 1-μg dose was inhibitory.

It is of interest that the effects of 5-fluoruracil, a common cancer chemotherapeutic agent, can be reversed with poly A:U.[122] Thus, parotid gland cells and pancreatic acinar cells, which secrete digestive enzymes, do so at a 200% higher level when 1 mg/k poly A:U was injected into Sprague-Dawley rats treated with this drug. Normal rats were not affected.

Lacour et al., have studied the effect of poly A:U on a mouse spontaneous mammary adenocarcinoma and a transplantable melanoma in hamsters.[123] In the former model, reduction of the tumor through surgery increased the mean survival time from 78 days to 102 days, whereas poly A:U given i.p. in 200-μg amounts every other day for 2 weeks following surgery increased the MST still further to 132 days. Comparable figures in the hamster model were 52 days for the control and 71 and 80 days for surgery alone and surgery plus poly A:U, respectively. Although the poly A:U effect on MST in the hamster was not dramatic, the number of metastases was reduced by 50% when poly A:U plus surgery was compared to surgery alone. These results, together with other studies, have served as a basis for recent initiation of a trial of poly A:U in over 100 human beings with cancer.[124]

A preliminary report on the effect of poly A:U introduced into cancer patients indicates some improvement in the anergic state of immunity without exerting toxicity.[125] In individuals with advanced cancer who were infused with 15 mg poly A:U twice a week for 3 to 24 weeks, 3 of 7 DNCB (−) patients were converted to DNCB (+). The in vitro PHA response was enhanced following poly A:U in one patient in which it was low initially, but this reaction was depressed by injection of poly A:U in four others. The latter finding is in keeping with previous evidence.[56,57] The in vitro response to normal antigens was enhanced in 3/7 patients. For obvious reasons, the doses used in this initial study were much less on a comparative-weight basis than those necessary for effectiveness in mice.

SUMMARY

The effective adjuvant action of the double-stranded synthetic polyribonucleotides, poly I:C and poly A:U, on both humoral and cell-mediated components of the immune response has been reviewed. The single-stranded polymers occasionally were reported as active in vitro, but in vivo required incorporation in Freund's incomplete adjuvant. Although the mode of action of these complexes is unknown as yet, their general characteristics, physicochemical requirements, relative toxicities, and cell types affected have been described. In addition, their applicability to aid in resistance to infectious diseases, immunologic diseases, and tumor immunity is discussed.

REFERENCES

1. **Braun, W. and Firshein, W.,** Biodynamic effects of oligonucleotides, *Bacteriol. Rev.,* 31, 83, 1967.
2. **Merritt, K. and Johnson, A. G.,** Studies on the adjuvant action of bacterial endotoxins on antibody formation. IV. Enhancement of antibody formation by nucleic acids, *J. Immunol.,* 94, 416, 1965.
3. **Johnson, A. G. and Hoekotra, G.,** Acceleration of the primary antibody response, in *Ontogeny of Immunity,* Smith, R. T., Good, R. A., and Miescher, P. A., Ed., University of Florida Press, Gainesville, 1967.
4. **Jaroslow, B. and Taliaferro, W.,** Restoration of hemolysin-forming capacity in x-irradiated rabbits by tissue and yeast preparations, *J. Infect. Dis.,* 98, 75, 1956.
5. **Taliaferro, W. H. and Jaroslow, B. N.,** The restoration of hemolysin formation in x-rayed rabbits by nucleic acid derivatives and antagonists of nucleic acid synthesis, *J. Infect. Dis.,* 107, 341, 1960.
6. **Simic, M. M., Petkovic, M., and Mancic, D.,** Partial restoration of hemolysin forming capacity in x-irradiated rats by homologus spleen disoxyribonucleic acid, *Bull. Inst. Nucl. Sci.,* 9, 155, 1959.
7. **Braun, W. and Nakano, M.,** Antibody formation: Stimulation by polyadenylic and polycytidylic acids, *Science,* 157, 819, 1967.
8. **Braun, W., Nakano, M., Jaraskova, L., Yajima, Y., and Jimenez, L.,** Stimulation of antibody forming cells by oligonucleotides of known composition, in *Nucleic Acids in Immunology,* Plescia, O. J. and Braun, W., Eds., Springer-Verlag, New York, 1968, 347.
9. **Johnson, A. G., Schmidtke, J., Merritt, K., and Han, I.,** Enhancement of antibody formation by nucleic acids and their derivatives, in *Nucleic Acids in Immunology,* Plescia, O. J. and Braun, W., Eds., Springer-Verlag, New York, 1968, 379.
10. **Braun, W., Ishizuka, M., Yajima, Y., Webb, D., and Winchurch, R.,** Spectrum and mode of action of poly A:U in the stimulation of immune responses, in *Biological Effects of Polynucleotides,* Beers, R. F. and Braun, W., Eds., Springer-Verlag, New York, 1971, 139.
11. **Johnson, A. G., Cone, R. E., Friedman, H. M., Han, I. H., Johnson, H. G., Schmidtke, J. R., and Stout, R. D,** Stimulation of the immune system by homopolyribonucleotides, in *Biological Effects of Polynucleotides,* Beers, R. F. and Braun, W., Eds., Springer-Verlag, New York, 1971, 157.
12. **Johnson, A. G.,** The adjuvant action of polynucleotides, *J. Reticuloendothel. Soc.,* 14, 441, 1973.
13. **Braun, W.,** RNA's as amplifiers of specific signals in immunity, *Ann. N.Y. Acad Sci.,* 207, 17, 1973.
14. **Han, I. H., Johnson, A. G., Cook, J., and Han, S. S.,** Comparative biological activity of endotoxin and synthetic polyribonucleotides, *J. Infect. Dis.,* 128, S232, 1973.
15. **Braun, W.,** Immunologic and antineoplastic effects of endotoxin: Role of membranes and mediation by cyclic adenosine-3′,5′-monophosphate, *J. Infect. Dis.,* 128, S188, 1973.
16. **Johnson, A. G. and Han, I. H.,** Mitogenic activity of polynucleotides on thymus influenced lymphocytes, in *Cyclic AMP, Cell Growth and the Immune Response,* Braun, W., Lichtenstein, L. M., and Parker, C. W., Eds., Springer-Verlag, New York, 1974, 44.
17. **Braun, W.,** Regulatory factors in the immune response: Analysis and perspective, in *Cyclic AMP, Cell Growth and the Immune Response,* Braun, W., Lichtenstein, L. M., and Parker, C. W., Eds., Springer-Verlag, New York, 1974, 4.
18. **Han, I. H. and Johnson, A. G.,** The mitogenic activity of polyadenylic-polyuridylic acid complexes, *Ann. N. Y. Acad. Sci.,* 249, 370, 1975.
19. **Johnson, A. G.,** Adjuvants to the immune system, *Annu. Rep. Med. Chem.,* 9, 244, 1974.
20. **Schmidtke, J. R. and Johnson, A. G.,** Regulations of the immune system by synthetic polynucleotides. I. Characterization of adjuvant action on antibody synthesis, *J. Immunol.,* 106, 1191, 1971.
21. **Menge, A. C.,** Use of polynucleotides with seminal antigens to induce isoantibodies and infertility in rabbits, *Proc. Soc. Exp. Biol. Med.,* 135, 108, 1970.
22. **Gillis, S. K., Tourtellotte, W. W., Mathews, K. P., Johnson, A. G., Murphy, W. H., and Abrams, G. D.,** Anaphylactic-like death provoked by polynucleotide adjuvants in the sensitized inbred mouse challenged intracerebrally, to be published.
23. **Braun, W. and Ishizuka, M.,** Antibody formation: Reduced responses after administration of excessive amounts of nonspecific stimulators, *Proc. Natl. Acad. Sci. U.S.A.,* 68, 1114, 1971.
24. **Ishizuka, M., Braun, W., and Matsumoto, T.,** Cyclic AMP and immune responses. I. Influence of poly A:U and cAMP on antibody formation in vitro, *J. Immunol.,* 107, 1027, 1971.
25. **Jaroslow, B. N. and Ortiz-ortiz, L.,** Influence of poly A-poly U on early events in the immune response in vitro, *Cell. Immunol.,* 3, 123, 1972.
26. **Bick, P. and Johnson, A. G.,** to be published.
27. **Jaroslow, B. N., Ortiz-ortiz, L., and Lenta, Sister Petra,** The adjuvant effect of natural and artificial RNA-polynucleotides in the immune response in culture, *Ann. N.Y. Acad. Sci.,* 207, 327, 1973.
28. **Braun, W. and Ishizuka, M.,** Cyclic AMP and immune responses. II. Phosphodiesterase inhibitors as potentiators of polynucleotide effects on antibody formation, *J. Immunol.,* 107, 1036, 1971.
29. **Ishizuka, M., Gafni, M., and Braun, W.,** Cyclic AMP effects on antibody formation and their similarities to hormone mediated events, *Proc. Soc. Exp. Biol. Med.,* 134, 963, 1970.

30. **Cone, R. E. and Marchalonis, J. J.,** Adjuvant action of poly A:U on T cells during the primary immune response in vitro, *Aust. J. Exp. Biol. Med. Sci.,* 50, 69, 1972.
31. **Stinebring, W. R. and Absher, M.,** The double-bitted axe: A study of toxicity of interferon releasers, in *Biological Effects of Polynucleotides,* Beers, R. F. and Braun, W., Eds., Springer-Verlag, New York, 1971, 249.
32. **Philips, F. S., Fleisher, M., Hamilton, L. D., Schwartz, M. K., and Sternberg, S. S.,** Polyinosinic-polycytidylic acid toxicity, in *Biological Effects of Polynucleotides,* Beers, R. F. and Braun, W., Eds., Springer-Verlag, New York, 1971, 259.
33. **Hilleman, M. R.,** Prospects for the use of double stranded ribonucleic acid (poly I:C) inducers in man, *J. Infect. Dis.,* 121, 196, 1970.
34. **deClercq, E., Wells, R. D., and Merigan, T. C.,** Increase in antiviral activity of polynucleotides by thermal activation, *Nature,* 226, 364, 1970.
35. **Niblack, J. F. and McCreary, M. B.,** Relationships of biological activities of poly I:poly C to homopolymer molecular weights, *Nat. New Biol.,* 233, 52, 1971.
36. **Schell, P. L.,** Uptake of polynucleotides by intact mammalian cells. VIII. Synthetic homoribopolynucleotides, *Biochim. Biophys. Acta,* 240, 472, 1971.
37. **Kelly, T. A. and Levy, H. B.,** Lack of uptake of intact polyriboinosinic-polyribocytidylic acid by cells, *Proc. Soc. Exp. Biol. Med.,* 144, 534, 1973.
38. **Schell, P. L. and Muller, W.,** Uptake of polynucleotides by intact mammalian cells, XII. Stimulation of the uptake of ribopolynucleotides by deoxyribonuclease, *Biochim. Biophys. Acta,* 247, 502, 1971.
39. **Schell, P. L.,** Uptake of polynucleotides by intact mammalian cells. XI. pH dependent permeability changes for synthetic ribopolymers, *Biochim. Biophys. Acta,* 262, 467, 1972.
40. **Schell, P. L.,** Uptake of polynucleotides by intact mammalian cells. XIV. Stimulation of uptake of polynucleotides by poly (L-lysine), *Biochim. Biophys. Acta,* 340, 323, 1974.
41. **Mayhew, E. and Juliano, R.,** Interaction of polynucleotides with cultured mammalian cells. II. Cell surface change density and RNA uptake, *Exp. Cell Res.,* 77, 409, 1973.
42. **Colby, C., Chamberlin, M. J., Duesberg, P. H., and Simon, M. T.,** The specificity of interferon induction, in *Biological Effects of Polynucleotides,* Beers, R. F. and Braun, W., Eds., Springer-Verlag, New York, 1971, 79.
42a. **Fenster, E. D., Lacour, F., and Harel, J.,** Nuclear penetration and stimulation of nucleic acids synthesis by poly(A)-Poly(U) in mammalian cells, *Exp. Cell Res.,* 94, 315, 1975.
43. **Collavo, D., Finco, B., and Chieco-Bianchi, L.,** Immune reactivity following poly I:poly C treatment in mice, *Nat. New Biol.,* 237, 154, 1972.
44. **Stout, R. D. and Johnson, A. G.,** Regulation of the immune system by synthetic polynucleotides. V. Effect on cell-associated immunoglobulin receptors and immunological memory, *J. Exp. Med.,* 135, 45, 1972.
45. **Stout, R. D. and Coons, A. H.,** Studies on membrane-bound receptors for antigen. Preparation of populations of receptor-depleted lymphocytes, *J. Exp. Med.,* 138, 875, 1973.
46. **Chester, T. J., deClercq, E., and Merigan, T. C.,** Effect of separate and combined injections of poly rI:poly rC and endotoxin on reticuloendothelial activity, interferon and antibody production in the mouse, *Infect. Immun.,* 3, 516, 1971.
47. **Berry, L. J., Smythe, D. S., Colwell, L. S., Shoengold, R. J., and Actor, P.,** Comparison of the effects of a synthetic polyribonucleotide with the effects of endotoxin on selected host responses, *Infect. Immun.,* 3, 444, 1971.
48. **Fruitstone, M. J., Shands, Jr., J. W., and Gifford, E. G.,** Biological effects of poly rI:rC and endotoxin in mice infected with *Mycobacterium* BCG: Interferon induction and toxicity, *Proc. Soc. Exp. Biol. Med.,* 139, 1104, 1972.
49. **Kong, Y. M. and Capanna, S. L.,** Interference with tolerance induction in mice by poly A:U, *Cell. Immunol.,* 11, 448, 1974.
50. **Capanna, S. L. and Kong, Y. M.,** Further studies on the prevention of tolerance induction by poly A:U, *Immunology,* 27, 647, 1974.
51. **Capanna, S. L. and Kong, Y. M.,** Effect of poly A:U on induction of T cell tolerance, *Fed. Proc.,* 34, 1029, 1975.
52. **Steinberg, A. D. and Talal, N.,** Suppression of antibody to nucleic acids with poly I:C and cyclophosphamide in murine lupus, *Clin. Exp. Immunol.,* 7, 687, 1970.
53. **Morton, J. I., Brown, M., and Siegel, B. V.,** Influence of cytoxan and poly I:C on early anti-nuclear antibody responses in New Zealand black mice, *Proc. Soc. Exp. Biol. Med.,* 139, 1417, 1972.
54. **Turner, W., Chan, S. P., and Chirigos, M. A.,** Stimulation of humoral and cellular antibody formation in mice by poly Ir:Cr, *Proc. Soc. Exp. Biol. Med.,* 133, 334, 1970.
55. **Cone, R. E. and Johnson, A. G.,** Regulation of the immune system by synthetic polynucleotides. II. Action on antigen-reaction cells of thymic origin, *J. Exp. Med.,* 133, 665, 1971.
56. **Friedman, H. M., Johnson, A. G., and Pan, P.,** Stimulatory effect of polynucleotides on short term leukocyte cultures, *Proc. Soc. Exp. Biol. Med.,* 132, 916, 1969.
57. **Chess, L., Levy, C., Schmukler, M., Smith, K., and Mardiney, M. R., Jr.,** The effect of synthetic polynucleotides on immunologically induced tritiated thymidine incorporation: Amplification of response, *Transplantation,* 14, 748, 1972.
58. **Cantor, H., Asofsky, R., and Levy, H. B.,** The effect of polyinosinic-poly-cytidylic acid upon grafts vs. host activity in Balb/c mice, *J. Immunol.,* 104, 1035, 1970.

59. **Kim, Y. G. and Han, S. S.,** Studies on hypoxia. X. Effects of synthetic polynucleotides on pancreata from hypoxic and control rats, an electron microscopic and biochemical study, *Exp. Mol. Pathol.,* 23, 43, 1975.
60. **McNeill, T. A.,** The effect of synthetic double stranded polyribonucleotides on haemopoietic colony forming cells in vitro, *Immunology,* 21, 741, 1971.
61. **McNeill, T. A. and Killen, M.,** The effect of synthetic double stranded polyribonucleotides on haemopoietic colony forming cells in vivo, *Immunology,* 21, 751, 1971.
62. **Scheid, M. P., Hoffmann, M. K., Komuro, K., Hammerling, U., Abbott, J., Boyse, E. A., Cohen, G. H., Hooper, J. A., Schulof, R. S., and Goldstein, A. L.,** Differentiation of T cells induced by preparations from thymus and by non-thymic agents, *J. Exp. Med.,* 138, 1027, 1973.
63. **Johnson, H. G. and Johnson, A. G.,** Regulation of the immune system by synthetic polynucleotides. II. Action on peritoneal exudate cells, *J. Exp. Med.,* 133, 649, 1971.
64a. **Alexander, P. and Evans, R.,** Endotoxin and double stranded RNA render macrophages cytotoxic, *Nat. New Biol.,* 232, 76, 1971.
64b. **Ma, B. I. and Johnson, A. G.,** Cytotoxic Activity of Macrophages Against Syngeneic Transplantable Leukemia, Line I_b in C58 Mice, Ph.D. thesis, University of Michigan, 1975, to be submitted.
65. **Cone, R. E. and Johnson, A. G.,** Regulation of the immune system by synthetic polynucleotides. IV. Amplification of proliferation of thymus influenced lymphocytes, *Cell. Immunol.,* 3, 283, 1972.
66. **Cone, R. E. and Wilson, J. D.,** Adjuvant action of poly A:U on T and B rosette-forming cells in SRBC immunized mice. Analysis by two rosette assay methods, *Int. Arch. Allergy Appl. Immuol.,* 43, 123, 1972.
67. **Hamaoka, T. and Katz, D. H.,** Mechanism of adjuvant activity of poly A:U on antibody responses to hapten-carrier conjugates, *Cell Immunol.,* 7, 246, 1973.
68. **Yamashita, U. and Kitagawa, M.,** Enhancing activity of synthetic polynucleotides on the induction of anti-hapten response, *Immunology,* 26, 925, 1974.
69. **Ruhl, H., Vogt, W., Bochert, G., and Diamantstein, T.,** Stimulation of B cells by poly A:poly U and poly I:poly C in vitro, *Immunology,* 26, 937, 1974.
70. **Scher, I., Strong, D. M., Ahmed, A., Knudsen, R. C., and Sell, K. W.,** Specific murine B-cell activation by synthetic single and double stranded polynucleotides, *J. Exp. Med.,* 138, 1545, 1973.
71. **Strong, D. M., Ahmed, A. A., Scher, I., Knudsen, R. C., and Sell, K. W.,** Specificity of in vitro murine B cell activation by protein and polysaccharide polymers, *J. Immunol.,* 113, 1429, 1974.
72. **Dutton, R. W.,** T cell factors in the regulation of the B cell response, in *The Immune System, Genes, Receptors, Signals,* Sercarz, E., Williamson, A. R., and Fox, C. F., Eds., Academic Press, New York, 1974, 485.
73. **Bick, P. and Johnson, A. G.,** Homopolyribonucleotide Induced Secretion of Immunocompetent Cell Enhancing Factors from Mouse Thymocytes, Ph.D. thesis, University of Michigan, 1975, to be published.
74. **Niederhuber, J.,** personal communication.
75. **Campbell, P. A. and Kind, P.,** Bone marrow-derived cells as target cells for polynucleotide adjuvants, *J. Immunol.,* 107, 1419, 1971.
76. **Mozes, E., Shearer, G. M., Sela, M., and Braun, W.,** Genetic control of immune responses to synthetic polypeptides in mice: Cellular analysis of the phenotypic correction of the Ir-3 gene defect by polyadenylic-poly-uridylic acid, *J. Immunol.,* 111, 439, 1973.
77. **Diamantstein, T., Wagner, B., Beyse, I., Odenwald, M. V., and Schulz, G.,** Stimulation of humoral antibody formation by polyanions. I. Effect of polyacrylic acid on the primary immune response in mice immunized with SRBC, *Eur. J. Immunol.,* 1, 335, 1971.
78. **Diamantstein, T., Wagner, B., Beyse, I., Odenwald, M. V., and Schulz, G.,** Stimulation of humoral antibody formation by polyanions. II. The influence of sulfate esters of polymers on the immune response in mice, *Eur. J. Immunol.,* 1, 340, 1971.
79. **Diamantstein, T., Wagner, B., Lage-Stehr, J., Beyse, I., Odenwald, M., and Schulz, G.,** Stimulation of humoral antibody formation by polyanions. III. Restoration of the immune response to SRBC by polyanions in Tx and lethally irradiated mice protected with bone marrow cells, *Eur. J. Immunol.,* 1, 302, 1971.
80. **Diamantstein, T., Wagner, B., Odenwald, M. V., and Schulz, G.,** Stimulation of humoral antibody formation by polyanions. IV. Effects of $DxSO_4$ on the kinetics of secondary immune responses in mice, *Eur. J. Immunol.,* 1, 426, 1971.
81. **Diamantstein, T., Meinhold, H., and Wagner, B.,** Stimulation of humoral antibody formation by polyanions. V. Relationship between enhancement of SRBC uptake by the spleen and adjuvant action of $DxSO_4$, *Eur. J. Immunol.,* 1, 429, 1971.
82. **Dean, J. H., Waleen, W. C., and Lucas, D. O.,** Polyinosinic-polycytidylic acid activation of mouse spleen cells in vitro, *Nat. New Biol.,* 237, 218, 1972.
83. **Hilleman, M. R., Lampson, G. P., Tytell, A. A., Field, A. K., Nemes, M. M., and Krakoff, I. H.,** Double stranded RNA's in relation to interferon induction and adjuvant activity, in *Biological Effects of Polynucleotides,* Beers, R. F. and Braun, W., Eds., Springer-Verlag, New York, 1971, 27.
84. **Woodhour, A. F., Friedman, A., Tytell, A. A., and Hilleman, M. R.,** Hyperpotentiation by synthetic double-stranded RNA of antibody responses to influenza virus vaccine in Adjuvant 65, *Proc. Soc. Exp. Biol. Med.,* 131, 809, 1969.
85. **Fenge, P. and Postic, B.,** Protection of rabbits against experimental rabies by poly I:C, *Nature,* 226, 171, 1970.

86. **Branche, R. and Renoux, G.,** Stimulation of rabies vaccine in mice by low doses of polyadenylic:polyuridylic complex, *Infect. Immun.,* 6, 324, 1972.
87. **Weinstein, M. J., Waitz, J. A., and Came, P. E.,** Induction of resistance to bacterial infections of mice with poly I:C, *Nature,* 226, 170, 1970.
88. **Renoux, G., Renoux, M., and Branche, R.,** Stimulation of antibacterial vaccination in mice by polyadenylic acid:polyuridylic complex, *Infect. Immun.,* 6, 699, 1972.
89. **Cheers, C. and Cone, R. E.,** Effect of polyadenine:polyuridine on brucellosis in conventional and congenitally athymic mice, *J. Immunol.,* 112, 1535, 1974.
90. **Fauve, R. M.,** Influence of synthetic polynucleotides on the bactericidal power of mouse macrophages against Listeria monocytogenes, in *Biological Effects of Polynucleotides,* Beers, R. F. and Braun, W., Eds., Springer-Verlag, New York, 1971, 215.
91. **Mackaness, G. B.,** Cell-mediated immunity to infection, in *Immunobiology,* Good, R. A. and Fisher, D. W., Eds., Sinauer Assoc., Stanford, Conn., 1971.
92. **Remington, J. S. and Merigan, T. C.,** Synthetic polyanions protect mice against intracellular bacterial infections, *Nature,* 226, 361, 1970.
93. **Youmans, A. S. and Youmans, G. P.,** Failure of synthetic polynucleotides to affect the immunogenicity of mycobacterial ribonucleic acid and ribosomal protein preparations, *Infect. Immun.,* 3, 149, 1971.
94. **Paul, B. B., Jacobs, A. A., Strauss, R. R., and Sbarra, A. J.,** The role of the phagocyte in host-parasite interactions. XX. Restoration of x-irradiation phagocytic damage by endotoxin or polyadenylic-polyuridylic acids, *J. Reticuloendothel. Soc.,* 7, 743, 1970.
95. **Cox, C. G. and Rafter, G. W.,** Pyrogen and enzyme release from rabbit blood leucocytes promoted by endotoxin and poly I:C, *Biochem. Med.,* 5, 227, 1971.
96. **Araujo, F. G. and Remington, J. S.,** Protection against *Toxoplasma gondii* in mice immunized with *Toxoplasma* cell fractions, RNA, and synthetic polyribonucleotides, *Immunology,* 27, 711, 1974.
97. **Lacour, F., Nahon-Merlin, E., and Michelson, M.,** Immunological recognition of polynucleotide structure, *Curr. Top. Microbiol. Immunol.,* 62, 1, 1973.
98. **Field, A. K., Tytell, A. A., Lampson, G. P., and Hilleman, M. R.,** Antigenicity of double-stranded ribonucleic acids, including poly I:C, *Proc. Soc. Exp. Biol. Med.,* 139, 1113, 1972.
99. **Van Boxel, J. A., Steinberg, A. D., and Green, I.,** A study of the cell-mediated and antibody responses to synthetic ribonucleic acids in guinea pigs, *J. Immunol.,* 108, 446, 1972.
100. **Chused, T. M., Steinberg, A. D., and Parker, L. M.,** Enhanced antibody response of mice to polyinosinic-polycytidylic acid by anti-thymocyte serum and its age-dependent loss in NZB-W mice, *J. Immunol.,* 111, 52, 1973.
101. **Parker, L. M. and Steinberg, A. D.,** The antibody response to polyinosinic-polycytidylic acid, *J. Immunol.,* 110, 742, 1973.
102. **Talal, N. and Steinberg, A. D.,** Immunity and tolerance to synthetic polynucleotides in New Zealand mice, in *Biological Effects of Polynucleotides,* Beers, R. F. and Braun, W., Eds., Springer-Verlag, New York, 1971, 231.
103. **Jacobs, M. D., Steinberg, A. D., Gordon, J. K., and Talal, N.,** Adjuvant effects of poly I:poly C in New Zealand mice, *Arthritis Rheum.,* 15, 201, 1972.
104. **Natali, P. G. and Tan, E. M.,** Precipitin reaction between polyribonucleotides and heat labile serum factors, *J. Immunol.,* 108, 318, 1972.
105. **Yachnin, S., Rosenblum, D., and Chatman, D.,** Biological properties of polynucleotides. VI. Further studies on the mechanism of complement inhibition by polyinosinic acid: The inactivation of the fourth component of C′, *J. Immunol.,* 93, 549, 1964.
106. **Gumbiner, C., Paterson, P. Y., Youmans, G. P., and Youmans, A. S.,** Adjuvanticity of mycobacterial RNA and poly A:U for induction of experimental allergic encephalomyelitis in guinea pigs, *J. Immunol.,* 110, 309, 1973.
107. **Paterson, P. Y. and Drobish, D. G.,** Adjuvanticity of synthetic polynucleotides for induction of experimental allergic encephalomyelitis in guinea pigs, *J. Immunol.,* 113, 1942, 1975.
108. **Casavant, C. H. and Youmans, G. P.,** The adjuvant activity of mycobacterial RNA preparations and synthetic polynucleotides for induction of delayed hypersensitivity to purified protein derivative in guinea pigs, *J. Immunol.,* 114, 1014, 1975.
109. **Casavant, C. H. and Youmans, G. P.,** The induction of delayed hypersensitivity in guinea pigs to poly U and poly A:U, *J. Immunol.,* 114, 1506, 1975.
110. **Winchurch, R. and Braun, W.,** Antibody formation: Premature initiation by endotoxin or synthetic polynucleotides in newborn mice, *Nature,* 223, 843, 1969.
111. **Han, I. H. and Johnson, A. G.,** Restoration of antibody formation in young and aged mice by poly A:U, to be submitted.
112. **Levy, H. B., Adamson, R., Carvone, P., De Vita, V., Gazdar, A., Rhim, J., Weinstein, A., and Riley, F.,** Studies on the anti-tumor action of poly I: poly C, in *Biological Effects of Polynucleotides,* Beers, R. F. and Braun, W., Eds., Springer-Verlag, New York, 1971, 55.
113. **Webb, D., Braun, W., and Plescia, O. J.,** Antitumor effects of polynucleotides and theophylline, *Cancer Res.,* 321, 814, 1972.

114. **Zeleznick, L. D. and Bhuyan, B. K.,** Treatment of leukemic (L 1210) mice with double stranded polynucleotides, *Proc. Soc. Exp. Biol. Med.,* 130, 126, 1969.

115. **Larson, V. M., Panteleakis, P. N., and Hilleman, M. R.,** Influence of synthetic double-stranded ribonucleic acid (poly I:C) on SV_{40} viral oncogenesis and transplant tumor in hamsters, *Proc. Soc. Exp. Biol. Med.,* 133, 14, 1970.

116. **Larson, V. M., Clark, W. R., and Hilleman, M. R.,** Influence of synthetic (poly I:C) and viral double-stranded ribonucleic acids on adenovirus 12 oncogenesis in hamsters, *Proc. Soc. Exp. Biol. Med.,* 131, 1002, 1969.

117. **Fisher, J. C., Cooperband, S. R., and Mannick, J. A.,** The effect of polyinosinic-polycytidylic acid on the immune response of mice to antigenically distinct tumors, *Cancer Res.,* 32, 889, 1972.

118. **Kreider, J. W. and Benjamin, S. A.,** Tumor immunity and the mechanism of polyinosinic-polycytidylic acid inhibition of tumor growth, *J. Natl. Cancer Inst.,* 49, 1303, 1972.

119. **Tennant, R. W., Hanna, Jr., M. G., and Farrelly, J. G.,** Effects of poly (2′-*O*-methyladenylic acid) on susceptibility and autogenous immunity to RNA tumor virus oncongensis in vivo, *Proc. Natl. Acad. Sci. U.S.A.,* 71, 3167, 1974.

120. **Fourcade, A., Friend, C., Lacour, F., and Holland, J. G.,** Protective effect of immunization with polyinosinic polycytidylic acid complexed with methylated bovine serum albumin against Friend leukemia virus in mice, *Cancer, Res.,* 34, 1749, 1974.

121. **Wagner, H. and Cone, R. E.,** Adjuvant effect of poly A:U upon T cell mediated in vitro cytotoxic allograft responses, *Cell. Immunol.,* 10, 394, 1974.

122. **Han, S. S.,** personal communication.

123. **Lacour, F., Spira, A., Lacour, J., and Prade, M.,** Polyadenylic-polyuridylic acid, an adjunct to surgery in the treatment of spontaneous mammary tumors in C3H-He mice and transplantable melanoma in the hamster, *Cancer Res.,* 32, 648, 1972.

124. **Lacour, J., Lacour, F., Spira, A., Delage, G., and Michelson, A. M.,** An immunotherapeutic trial of the adjuvant poly A:U in the management of breast cancer, *Bull. Cancer* (Paris), 61, 275, 1974.

125. **Wanebo, H. J., Oettgen, H. F., Lundy, J., Stock, C. C., and Old, L. J.,** Influence of poly A:U on immune responses in cancer patients, *Am. Assoc. Cancer Res.,* Abstr. 716, 1975.

DIALYZABLE AND NONDIALYZABLE TRANSFER FACTOR

Peter Baram

TABLE OF CONTENTS

INTRODUCTION

The transfer of delayed-type hypersensitivity (DH) from picryl chloride-sensitive to nonsensitive guinea pigs using intact white blood cells was first demonstrated by Landsteiner and Chase.[1] Viable cells appeared to be essential for the transfer of DH. Killed guinea pig leukocytes lost the capacity to transfer DH.[2] Jeter et al.[3] subsequently reported that lysates prepared from 2,4-dinitrochlorobenzene-sensitive guinea pig peritoneal exudate cells were capable of transferring donor-specific hypersensitivity to unsensitized guinea pigs. This sensitive state was verified by skin testing the recipient animals. The transfer in the guinea pig system with the white blood cell lysates was accomplished with great difficulty, and many investigators were unable to duplicate these observations. Lawrence[4] repeated the transfer of DH with intact leukocytes using human donors and recipients. Blood leukocytes obtained from human donors, sensitive to purified protein derivative (PPD), specifically transferred PPD-DH to nonsensitive human recipients. The recipients, skin test negative to PPD prior to receiving the leukocytes, developed DH to PPD following the transfer. Lawrence[5] then demonstrated that lysates prepared from human peripheral white blood cells transferred donor-specific DH to human recipients.

Lawrence et al.[6] subsequently demonstrated that transfer capacity could be found in the dialysate when the white blood cell lysates were dialyzed against distilled water. He referred to this biologically active preparation as dialyzable transfer factor (TF_d). Perhaps the three most unique features of TF_d are

1. The apparent capacity to transfer donor-specific hypersensitivity (this will be discussed in greater detail later) with low molecular weight material.
2. The persistence of the hypersensitive state in the normal recipient for several months or years.
3. The temporary reconstitution of immunologic competence in some immunodeficiency states.

The transfer capacity of the human leukocyte dialysate was verified by Baram and Mosko.[7] They chromatographed the TF_d on Sephadex G-25 and found transfer activity in a fraction of <10,000 mol wt. In addition, Baram and Mosko[7] and Baram et al.[8] reported that the nondialyzable portion of the human white blood cell lysate also had the capacity to transfer donor-specific DH to human recipients. They referred to this preparation as nondialyzable transfer factor (TF_{nd}); it contained components ranging in molecular weight from 70,000 to 300,000. TF_{nd} recipients also maintained the transferred hypersensitive state for several months.

There have been no studies to indicate the comparable effectiveness of TF_{nd} to TF_d in the transfer of DH or the reconstitution of immunocompetence in immunodeficient states. Early reported transfer studies with leukocytes lysates (TF_L) were, in reality, performed with mixtures of both TF_d and TF_{nd}. More recent studies have distinguished between TF_d, TF_{nd}, and TF_L. It is important to do so since the properties of TF_d and TF_{nd} may differ.

Potentially, TF preparations appear to have value both as immunotherapeutic agents in humans and as tools for investigating the mechanisms of the induction and expression of CMI at cellular and subcellular levels. Many investigations are currently in progress on the clinical value of TF_d in relation to tumor immunotherapy, treatment of chronic microbial or viral infections, reconstitution of immunodeficient patients, treatment of autoimmune diseases, and vaccine substitution. The reports that TF_d induces temporary immunologic competence and/or clinical improvement in recipients having a variety of distinct diseases suggest that the response to TF_d may be very complex and may be due to distinct components in the TF_d preparation. The transfer of specific hypersensitivity may be one of a number of host responses to TF_d. Adjuvant action,[9,10] chemotaxis,[11] elevation of the number of T-lymphocytes,[12] and other effects have also been reported in TF_d recipients.

TF_d and TF_{nd} are also being studied in a number of laboratories in attempts to determine the following.

1. Composition of TF preparations having distinct specificities.
2. The role of antigen in the DH transfer capacity of TF_d and TF_{nd}.
3. The relationship, if any, between TF_d, TF_{nd}, and (immunogenic) RNA.
4. The specificity of TF_d and TF_{nd} in transferring donor hypersensitivity.
5. Nonspecific effects of TF_d and TF_{nd} on the recipient reticuloendothelial system.
6. The cell source of TF_d and TF_{nd}.
7. The mechanism by which TF_d and TF_{nd} recruit lymphocytes and the types of lymphocytes recruited.

CHARACTERIZATION OF TF

Definition

The historical development of TF studies has led to a broadening of the original definition of TF and a plethora of studies yielding data which may or may not reflect TF activity. The original definition of TF_d and TF_{nd} specified that human peripheral white blood cell dialyzable or nondialyzable components transferred donor-specific DH to normal human recipients. A number of questions should be considered in evaluating TF studies: Should leukocyte lysates, dialysates, and nondialyzable components that do not transfer donor-specific DH, but have other effects on host immunocompetence, be referred to as TF? Do different preparations of TF vary in composition and biologic activity? Is DH sufficiently well understood so that a preparation that transfers donor hypersensitivity to a recipient, verified only by in vitro assay, may be called TF? What is the difference, if any, in using cells derived from different tissues and processed in a variety of ways on the TF product? In addition to the transfer of donor hypersensitivity, TF preparations appear to induce other responses by recipients. Some investigators have concerned themselves with effects that augment recipient CMI responses or reconstitute immune competence. These effects may be entirely unrelated to the transfer of specific DH to recipients. Clinical improvement in a variety of disease conditions has been equated with TF_d activity. The improvement of the patients' condition is used to evaluate the effectiveness of TF_d, and failure to demonstrate clinical improvement is assumed to indicate that TF_d is ineffective. Care must be taken in separating the transfer capacity of TF preparations from other biologic effects when evaluating TF therapy. Indeed, it is possible to effect transfer of donor DH without necessarily improving the condition of the patient.[13]

In attempting to establish in vitro models for demonstrating the specific transfer of donor DH, investigators have, for the most part, assumed that the TF_d and TF_{nd} components which have been responsible for in vivo transfers are also able to transfer in vitro by recruiting lymphocytes. No such evidence exists. The difficulty which investigators have had in establishing in vitro models has resulted in many studies which are directed to the augmentation of responses by sensitive lymphocytes in vitro in the presence of TF_d.[14] That such an effect is due to the same moities which transfer donor-specific DH in vivo must still be proven.

TF_d and TF_{nd} are terms also used to designate comparable white blood cell preparations from many species other than human which can transfer donor-specific hypersensitivity. In some species, particularly nonhuman primates, it is difficult to verify transfer of donor DH by skin testing the recipient. Skin test responses are very poor in many of these species. In vitro assays, such as lymphocyte transformation (LT) (usually determined as increased incorporation of tritiated thymidine) or the production of macrophage migration inhibitory factors (MIF) by lymphocytes from recipient animals, have, therefore, been used as substitutes for DH skin testing. Although these assays are considered to be in vitro correlates of CMI, they are dissociable from DH skin tests and under some conditions may reflect B-lymphocyte responses.[15]

Preparation of TF_d and TF_{nd}

TF composition and biologic activity may well depend on the level of donor DH, the tissue from which the cells have been obtained, and the method of preparation. Peripheral white blood cells, peritoneal exudate cells, lymph node cells, spleen cells, and alveolar cells have been used for the preparation of TF.[5,16,17] None of these cell preparations has been completely fractionated to yield pure populations of T- or B-lymphocytes or other leukocytes. Therefore, individual preparations of TF_d and TF_{nd} may vary considerably in composition and biologic activity, depending on the variety of cells and their percent contribution to the total lysate. More recently, investigators preparing TF_d from human leukocytes have used continuous cell separators for collecting cells or have enriched the mononuclear cell preparation by ficoll-hypaque density gradient centrifugation of blood. Since these procedures favor the accumulation of large numbers of mononuclear cells, it is possible that TF_d and TF_{nd} may be lymphocyte products;[18,19] however, this has not yet been determined.

The procedures used to prepare TF_d and TF_{nd} are usually those originally formulated by Lawrence[5] using peripheral blood leukocytes. Peripheral blood is collected into any one of a number of anticoagulants, heparin, ACD solution, or ethylenediaminetetracetate. The leukocytes are separated from the red blood cells and plasma. Spleen and lymph nodes are usually processed to produce single cell suspensions. The cells are washed and then disrupted. It is important that the cells be processed quickly. Many investigators have observed that lysates prepared from cells allowed to stand for several hours before disruption do not have any TF activity. Washed centrifuged cell pellets may be disrupted by freezing at −70°C and thawing at 37°C. Usually, 6 to 10 cycles are required to completely lyse the cell pellet. Alternate methods of cell disruption include sonication or N_2 decompression. The lysates may be treated with DNAse at this point.[5] DNAse treatment decreases the viscosity of the lysate, but it is not essential for the production of TF_d and TF_{nd}.[7,8]

Specific TF_d may be released from leukocyte suspensions by incubating the leukocytes for approximately 1 hr at 37°C with antigen to which the donor was sensitive. Lawrence and Papenheiner[20,21] reported that leukocytes sensitive to two antigens released TF only to the antigen with which they were incubated. Leukocytes obtained from PPD- and diphtheria toxoid (DT)-sensitive donors released into the medium TF_d specific for PPD when incubated with PPD. The leukocytes were then disrupted, and the TF_I obtained transferred moderate DT-DH to human recipients. The TF_I prepared from the disrupted cells did not transfer PPD sensitivity. The reciprocal experiments were not done. It would have been desirable to know if incubation of the intact leukocytes with DT resulted in the release of DT-TF into the medium while PPD-TF remained in the leukocytes. Graybill et al.[22] incubated peripheral blood leukocytes, obtained from coccidiodin skin-test-positive human donors, with coccidiodin for 1 hr. The medium was subsequently dialyzed, and the dialysates were used to treat patients with disseminated coccidiodomycosis. Two patients, anergic by skin test to coccidiodin before receiving the dialysate, developed positive DH skin test to coccidiodin following treatment. TF prepared in this manner contains dialyzable antigen fragments. These fragments may, in conjunction with cell components released from the leukocytes, cause active sensitization of the recipient. Oliveria-Lima[23] also confirmed that TF activity was released into the medium in which human leukocytes were incubated with antigen to which the donor was sensitive. This has been confirmed by others.[24]

TF_d is also released from leukocytes without specific antigen being present. This phenomenon may be useful in preparing TF_d. Burger and Jeter[16,25] reported that medium, in which peritoneal exudate and lymph node cells from 2,4-dinitrochlorobenzene (DNCB)-sensitive guinea pigs were incubated for 4 hr at 37°C, was capable of transferring DNCB sensitivity to nonsensitive guinea pigs. It was not necessary to add specific antigen to the incubating leukocytes. Neither Graybill et al.[22] nor Burger et al.[16,25] examined the medium for TF_{nd}. It appears that TF activity is released from leukocytes allowed to incubate in culture for several hours. The addition of specific antigen, to which the leukocyte donor was sensitive, may facilitate TF release, although the evidence is not conclusive.

Sonication of leukocytes has also been used to prepare TF_d. This procedure has not been used frequently. The TF_d moiety at least is sufficiently stable that many innovative procedures may be applicable to preparing TF.

Quantitation of TF_d and TF_{nd}

Attempts to define and quantitate a unit of TF_d or TF_{nd} have resulted in a variety of definitions. Investigators have used one or more of the following criteria to define a unit of TF_d or TF_{nd}.

1. The TF obtained from the total number of leukocytes derived from 500 ml of blood.
2. The amount of TF obtained from a preparation of white blood cells, spleen, peritoneal exudate, or lymph node cells having a known number of mononuclear cells.
3. A quantity of orcinol reactive material obtained from leukocytes (reflecting possible RNA concentration).
4. A quantity of protein obtained from the leukocytes.

Difficulties in defining a unit of TF_d or TF_{nd} arise from an inability to define the level of DH required of the TF cell donor and agreement on the biologic activity to assay, as well as the lack of a reliable and reproducible in vitro assay.

Where it is the intent of the investigator to use TF_d or TF_{nd} to transfer donor-specific hypersensitivity, it is important that the donor be exquisitely sensitive to the antigen. In the human system, for example, transfer of PPD hypersensitivity is most readily accomplished using TF_d or TF_{nd} prepared from the peripheral blood leukocytes of donors having a positive skin reaction of 50 mm or greater of induration to 0.0002 mg of PPD. Leukocytes obtained from weak skin reactors (having 10 to 20 mm of induration) are not a good source of TF_d or TF_{nd} that is to be used for the transfer of donor-specific hypersensitivity. The maximum skin reactivity of the donor to an antigen varies with the species; DH skin reactions as large as those found in exquisitely sensitive humans are usually not observed in other species. Demonstrating transfer of donor hypersensitivity may, therefore, be more difficult in other species, but is still possible. Other biologic effects observed with TF_d or TF_{nd} may be quite independent of donor hypersensitivity.

Although TF_d and TF_{nd} appear to be relatively stable, the types of cells and the manipulation involved in the production of TF may result in considerable variation in composition or biological activity. Huard and Baram[9] compared five lots of human TF_d for protein concentration, orcinol-

TABLE 1

Ribose and Protein Content of TF_d Pools[a]

TF_d pool	Ribose mg	Protein mg	260:280
Human			
H-1	2.025	22.080	N.D.
H-2	0.650	31.750	1.6
S-13	0.900	5.925	2.6
CY-1	0.285	3.150	2.5
CY-2	0.690	3.300	2.6

[a]Ribose, protein concentration and 260:280 absorbance ratio of five pools of human TF_d. Each pool was prepared from 5×10^8 lymphocytes.[9]

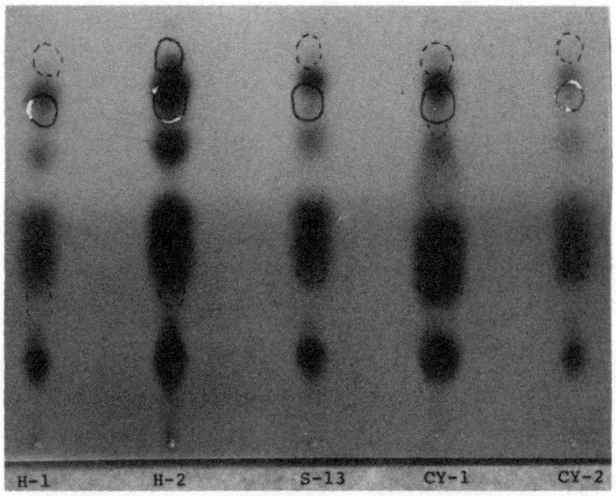

FIGURE 1. Silica gel thin-layer chromatography of five human preparations of TF_d The samples were diluted to represent 5×10^8 lymphocytes. The solvent consisted of butanol, acetane, acetate, and water. (From Huard, T. K. and Baram, P., in *Transfer Factor: Basic Properties and Clinical Application,* Ascher, M. S., Kirkpatrick, C. H., and Gottlieb, A. A., Eds., Academic Press, New York, 1976. With permission.)

reacting material, absorption at 260 and 280 nm, and composition using silica gel thin-layer chromatography. The samples were prepared from 5×10^8 lymphocytes regardless of the number of other leukocytes present. The blood was collected either in heparin, and the red blood cells were sedimented with dextran or, the leukocytes were collected on a cell separator. The leukocytes were separated from the plasma by centrifugation, and they were then washed once in saline. The cells were frozen and thawed, and the lysates were dialyzed. The dialysates were lyophilized and reconstituted in a small volume of water.

The composition of the five pools differed greatly (Table 1). Protein concentration ranged from 3.3 to 31.8 mg. The concentration of orcinol-reactive material varied from 0.29 to 2.03 mg. On thin-layer silica gel chromatography (Figure 1), the samples of the five pools were observed to have the same number of components. A minimum of ten components was detected. However, there were obvious differences in the intensity of staining with ninhydrin that did not reflect the concentration of protein (Table 1).

These variations in TF_d compositions may be responsible for the difference in the observed effectiveness of separate lots of TF_d used by investigators. Standardization will be more feasible if purification studies make possible the use of fractions with distinct biologic activities.

Early attempts to quantitate TF_L in the human TF model were based on the number of donor blood leukocytes needed to produce a lysate which could transfer local or total body sensitivity to the normal recipient. Lawrence[5] was able to demonstrate total body DH in normal human recipients using TF_L prepared from 8.5×10^7 blood leukocytes. Many of these recipients remained hypersensitive for several months. However, the number of leukocytes or lymphocytes necessary to transfer donor-specific DH depends on the degree of DH of the donor and the immunocompetence of the recipients. In addition, where effects are desired other than the transfer of specific DH, the number of leukocytes or lymphocytes needed may be different. Investigators have variously defined the unit of TF_d as that amount of dialysate obtained from 5×10^8 to 6×10^9 leukocytes. Transfer of DH is not always accomplished with a single or even multiple injection.

In transferring DNCB DH in the guinea pig, Burger and Jeter[16] and Burger et al.[25] found it necessary to use TF_d prepared from a pool of 1×10^9 peritoneal, lymph node, and alveolar washing cells in order to effect transfer in the normal guinea pig.

Maddison et al.[26] transferred DH to PPD from sensitive to nonsensitive rhesus monkeys using either TF_d or TF_{nd}. The investigators found that larger numbers of leukocytes were necessary to prepare the TF_d and TF_{nd} in order to effect a transfer of DH in the rhesus monkey than were required for human transfer: 3.4×10^9 lymphocytes. Condoulis and Baram[71] have, however, found that TF_d derived from approximately 5×10^8 leukocytes from keyhole limpet hemocyanin (KLH)-sensitive rhesus monkeys could transfer KLH sensitivity to two of four animals.

In preparing and assaying TF_d or TF_{nd} it is

important to establish a definition of the biologic activity to be investigated, the conditions under which the preparations are to be used, and the reliability of the assay for evaluating the biologic effect. Until TF_d and TF_{nd} are better characterized, it is likely that investigators will continue to define a unit of TF_d or TF_{nd} as that amount which in their hands is most capable of producing a desired effect in vivo.

MODEL SYSTEMS FOR STUDYING TRANSFER FACTOR

A. Human In Vivo Systems

Several models have been developed for studying TF_d and TF_{nd}. In some respects the human model is both the ideal and the least desirable model. Transfer of specific donor DH in the human system is more readily and consistently accomplished than in any other species model reported. This may be due to the unusually strong donor DH to naturally occurring antigens (as determined by skin testing) most often studied in the human TF system. DH skin reactivity in humans is greater than that obtained in other species. It is possible that DH is more readily detectable in humans than in other species following the injection of TF_d or TF_{nd} because of the easier perceptivity of the skin test reaction in the human. The level of sensitivity obtained following TF injection may be equivalent in different species. Is the difference observed due to the TF preparation, species receptivity, or skin reactivity? The more frequent use recently of in vitro assays for detecting CMI in transfer recipients makes it possible to evaluate and compare the human transfer factor model to other species models in which skin reactivity is not as strong as in the human.

An additional advantage of using the human model in TF studies is that the effect of TF in a variety of immunodeficiency states can be investigated. Since the ultimate aim would be to use TF in the treatment of human disease, the human is the ideal model for study. It is important to consider that in vitro assays for CMI are only correlates of DH. In some immunodeficiency states it is possible to demonstrate DH in the absence of positive in vitro correlate assays for CMI. Further, the in vitro correlate assays of CMI often do not correlate with each other.

There are many problems associated with the human TF model.

1. Studies require large numbers of human subjects, making experiments difficult to design adequately. They are costly to perform (this is particularly true in studies in which fractions of TF_d are evaluated for biologic activity).

2. Diet, alcohol consumption, medication, and intercurrent infections, as well as other factors, may affect the recipients' response to TF preparations.

3. From a practical point of view, it is usually easier to use TF leukocyte donors that are sensitive to naturally occurring antigens. (This has caused problems in evaluating the specificity of the recipient response. It is difficult to establish that recipients have not previously experienced the naturally occuring antigens. Therefore, experiments must be designed to distinguish between the transfer of donor hypersensitivity and the specific or nonspecific augmentation of recipient hypersensitivity.)

4. The protection of human subjects is a major problem (no long-term retrospective studies have been done to evaluate the effect of TF on normal recipients).

Two types of human transfer factor models have been used. Although leukocyte donors for TF have primarily been "normal" immunocompetent individuals, recipients have been either normal individuals or immunologically compromised patients with other associated disease processes. It is essential to distinguish between the transfer of donor DH to immunocompetent recipients and the transfer to recipients who are immunologically compromised. Potential recipients who are immunocompetent and are negative to the antigen specificity of the TF by in vivo and in vitro tests are usually presumed not to be sensitive to that antigen. However, such individuals may have experienced the antigen previously and have low levels of sensitivity or memory cells for it. The injection of TF may result in: (1) the transfer of specific hypersensitivity, (2) specific augmentation of a low level of sensitivity, and (3) nonspecific adjuvantlike effects for antigen used to skin test the recipient. With adequate controls, the distinction can be made among the three types of responses. Unfortunately, not all transfer studies have been designed to distinguish these effects. In the potential recipient who has an immunodeficiency, abnormalities may be due to a defect at any number of levels. The recipient may indeed have had previous exposure to the antigen for which the TF has specificity and may be sensitive to it, but the defect prevents expression of the response to the antigen. Unless the TF preparation has the properties necessary to correct the patients' defect (properties unrelated to the transfer of donor DH), the expression of DH in the TF recipient cannot be observed.

The heterogeneity of immunodeficiency states and the resultant patient response to TF_d have been observed. For example, patients with chronic mucocutaneous candidiasis can be placed into three classes.

1. Immunologically compromised with detectable CMI defects.
2. Defects apparently related only to candida hypersensitivity (anergy).
3. No detectable immunologic abnormality.

This categorization may be an oversimplification. Some patients may have negative DH skin test to candida antigen prior to TF_d treatment. However, lymphocyte response in vitro to candida antigen may be positive. In other patients with chronic candidiasis, DH, MIF, and lymphocyte incorporation of tritiated thymidine may all vary independently.[18,27,28] This again indicates the heterogeneity of the immune defects and the difficulties in evaluating TF_d studies on poorly defined immunodeficient patients. It is, therefore, preferable to treat the human as at least two distinct models: (1) transfer to immunocompetent humans and (2) transfer to a heterogeneous population of immunologically compromised humans. In the latter group the immunologic defect should be carefully characterized before TF studies are initiated.

Transfer to Immunocompetent Humans

Most of the reports demonstrating specific transfer to immunocompetent humans have been to naturally occurring antigens such as streptococcal M substance, PPD, coccidioidin, and diphtheria toxoid. Transferse were initially performed using leukocyte lysates (TF_L). Recipients were injected with TF_L and challenged intradermally 48 hr later with antigen to which the leukocyte donors were sensitive. Recipients developed typical DH skin test reactions. Normal recipients remained sensitive to the antigen from several months to more than 2 years. The long duration of the hypersensitive state indicated that the recipients were actively sensitized. Subsequently, Lawrence found transfer capacity to reside in a dialyzable component.[5] This observation was confirmed by Baram and Mosko and extended to describe TF_{nd}.[7]

As specificity controls in transfer experiments, TF_{nd} or TF_d prepared from the leukocytes of humans not sensitive to a specific antigen is injected into other recipients. Recipients receiving these preparations usually do not give positive DH skin test reactions when subsequently challenged with the antigen. However, negative control transfers have, on occasion, resulted in a positive DH skin test response by the recipient to an antigen to which neither the donor nor the recipient showed any prior detectable sensitivity. Such aberrations may be due to the adjuvant properties of the TF preparation, resulting in the elevation of low levels of sensitivity or in facilitating sensitization by the skin testing antigens.

In an interesting series of experiments designed to rule out the adjuvant effects of TF, Lawrence[29] transferred hypersensitivity to HL-A antigens. In the experimental design, individual B received four successive skin grafts from donor A. TF_L prepared from the leukocytes of B, following this sensitization, was able to induce total body sensitization in recipient C. Grafts from A when placed on C, after he had received the TF_L, were rejected as if they were second-set grafts. The rejection time was 4 to 5 days. A skin graft from an unrelated donor placed on C at the same time was rejected as a first-set graft at approximately 7 to 13 days. This would indicate that the rejection of the A skin graft was due to the transfer of hypersensitivity against HL-A antigens. It would have been valuable to have all of the recipients and donors HL-A typed. Although the transfer of the HL-A sensitivity involved natural antigens, it is difficult to determine if recipients may have experienced these antigens at an earlier date.

In order to resolve the problem of specific transfers or augmentation of existing sensitivity due to recipient contact with antigen prior to TF injection, investigators have attempted to transfer sensitivity to: (1) modified naturally occuring antigens, (2) antigens with geographically limited distribution, or (3) synthetic antigens. Maurer[30] prepared TF_L from the leukocytes of donors sensitized to ethylene oxide-treated human serum albumin (EO-HSA). He transferred EO-HSA DH to six recipients. After 1 year the recipients were still skin test positive to EO-HSA. Unfortunately, only one recipient received TF_L prepared from the leukocytes of a donor not sensitive to EO-HSA. The use of this antigen does not resolve the problem of recipient-antigen contact prior to TF injection since the recipient may have been responding to unaltered but newly exposed antigens.

Zukerman et al.[31] transferred KLH sensitivity with TF_d obtained from donors sensitized to KLH. The recipients were not skin tested with KLH before receiving the KLH-TF_d. Recipients receiving PPD-TF_d did not develop DH to KLH. This would indicate that there was specificity in the KLH-TF_d transfer. It does not, however, demonstrate that TF_d can sensitize recipients to antigens to which they are virgin. Lymphocytes from about 5% of unsensitized humans are stimulated to incorporate tritiated thymidine (^{3}H-T) by KLH. This may be a mitogenic effect or it may reflect a wide distribution of the KLH determinants. The KLH-TF_d may augment recipient sensitivity to KLH.

Transferring DH to DNCB with DNCB-TF_L to unprimed recipients has been unsuccessful. Brandiss[32] reported on attempts to transfer DNCB sensitivity to recipients having no known contact with DNCB; the same level of skin test reactivity was obtained in recipients receiving DNCB-TF_L or nonspecific TF_L when they were challenged with DNCB. Arala-Chaves and Pinto[33] were also unable to transfer DNCB hypersensitivity using DNCB-TF_L. However, if the recipients were primed by skin testing with DNCB before they were injected with DNCB-TF_L, they developed hypersensitivity to DNCB as demonstrated by a subsequent skin test. The studies on the transfer of DH to DNCB with TF are extremely important. They raise the question of the ability to transfer DH to antigens when the recipient is virgin to them. The transfer of DH cannot be accomplished with equanimity to all antigens. Transfer of DNCB-DH may be difficult to accomplish and not related to the need for prior exposure of the recipient to antigen. The work of Burger and Jeter[16] in the guinea pig model would indicate that prior exposure to antigen is not necessary in order to accomplish transfer with DNCB-TF.

Although Lawrence has reported that TF activity resided in the dialyzable portion of the leukocyte lysate, Baram and Mosko[7] and Baram et al.[8] found that it was also possible to transfer DH to PPD using the nondialyzable portion of the leukocyte lysate. PPD-TF_{nd} recipients developed DH skin test reactions upon challenge with PPD. The hypersensitive state was detectable for up to 6 months. Recipients receiving nonspecific TF_{nd}, prepared from the leukocytes of donors not sensitive to PPD, did not develop DH reactions when challenged with PPD. It has been suggested that the PPD transfer capacity of the TF_{nd} may have resulted from incomplete dialysis of TF_d. However, the authors reported that the lysates were dialyzed against several hundred volumes of distilled water for 3 days with changes at 24-hr intervals. It is unlikely that under these conditions dialysis was incomplete. The existence of a TF_{nd} as well as a TF_d raises the fundamental question of the relationship between the two components. Is TF_d a breakdown product of TF_{nd} or are they both breakdown products of larger molecules, produced by enzyme hydrolysis during leukocyte disruption? Are they distinct entities? There is a paucity of information in the human model with TF_{nd}. Concern about the transfer of portions of viral genomes with the higher molecular weight TF_{nd} may have limited investigations with TF_{nd}.

Transfer of DH and Alteration of CMI in Immunologically Compromised Humans

Transfer of DH or modification of immunodeficiency syndromes has been reported in five types of immunological abnormalities: (1) inadequate CMI in response to microorganisms causing chronic infectious diseases, (2) genetically inherited immunologic deficiency states, (3) immunodeficiencies of undetermined origin, (4) autoimmune diseases, and (5) unresolved tumors. As indicated earlier, it is difficult to ascertain if TF_d causes the same kind of response in immunologically compromised patients as in normal individuals. The transfer of specific donor hypersensitivity may be of little importance. Other "nonspecific" effects of leukocyte dialysate may be responsible for the effects observed in immunologically compromised patients. Investigators treating immunologically compromised patients must recognize and monitor at least three distinct effects which TF_d may have on the patients: (1) transfer of donor DH, (2) reconstitution of recipient impaired immunocompetence, and (3) alteration of the clinical condition of the patient.

The response to TF_d by compromised patients differs from the response of normal individuals: (1) donor DH is not transferred with the same frequency, (2) the transferred hypersensitive state may persist only for a few weeks to a few months, and (3) in vitro CMI assays and in vivo skin test responses to specific antigen may not correlate with each other.

Patients receiving coccidioidin-TF_d or candida-TF_d may subsequently develop positive DH reactions upon challenge with the specific antigen, but the LT and/or the MIF assay are discordant.[18,28,34-36] Kirkpatrick et al.[18] reported on a group of candidiasis patients who before receiving TF_d were anergic to candida and other antigens by skin test, lymphocyte transformation, and indirect MIF assays. Many of the patients did not respond to attempts to sensitize them to DNCB. Treatment of patients with candida-TF_d usually resulted in the development of DH to antigens to which the TF_d cell donor was sensitive. Such reactivity persisted for 1 week to 60 days. MIF production by the patient lymphocytes stimulated with candida antigen also became detectable after TF_d therapy. However, in vitro LT induced by candida antigen after TF_d therapy did not correlate well with skin test results or MIF production (Table 2). Conversion of skin test reactivity, MIF production, and LT do not have a consistently high correlation, and the patients do not always show clinical improvement. The immunologic defect is not always correctable by TF_d therapy. It is possible that the amount of TF_d given is inadequate for some patients.[37] The individual batches of TF_d should perhaps be assayed for biologic activity on animal models to achieve some degree of standardization. Transferring specific hypersensitivity to patients who are anergic may temporarily cause undesired effects. Bullock et al.[38] treated lepromatous leprosy patients with TF_d derived from the leukocytes of donors sensitive to *M. leprae* antigens. Four patients developed erythema nodosum and three developed fever and arthralgia. Hastings et al.,[39] however, found marked clinical improvement in leprosy patients treated with TF_d.

Tumor patients also have been demonstrated to have impaired CMI. Reports indicate that TF_d may be beneficial as an immunotherapeutic agent. Again, unlike normal recipients who remain sensitive for extended periods after receiving TF_d, tumor patients must be treated periodically with TF_d. The quantity and frequency of TF_d administered and the clinical response obtained usually vary with investigators.

Levin et al.[40] reported on a series of osteogenic sarcoma patients treated with TF_d. Following removal of the tumor, patient lymphocytes were found to be less cytotoxic for tumor cells. TF_d derived from 1×10^9 peripheral blood leukocytes of tumor antigen-sensitive donors was injected into patients twice weekly. Treatment with tumor-specific TF_d resulted in an elevation of lymphocytotoxicity to tumor cells in vitro. Nonspecific TF_d injections depressed the patient's lymphocytotoxicity response. Eight patients in this series were stabilized by tumor antigen-specific TF_d. The condition of four continued to deteriorate.

Price et al.[41] treated six patients having malignant melanoma with TF_d. Patients received TF_d obtained from 1×10^9 lymphocytes twice weekly for 3 weeks. The TF_d was prepared from the

TABLE 2

Summary of Immunologic Responses by Candidiasis Patients Before and After Transfer Factor

	Before transfer factor				After transfer factor			
	Skin tests				Skin tests			
Patient	Candida	CDNB	Lymphocyte transformation	MIF production	Candida	DNCB	Lymphocyte transformation	MIF production
R.H.	Neg[a]	Neg	Pos	Neg	Pos[a]	Neg	Pos	Pos
J.C.	Neg	Neg	Neg	Neg	Pos	Neg	Intermittent pos	Pos
R.V.	Neg	Neg	Neg	Neg	Pos	Neg	Neg	Pos
D.L.	Neg	Equivocal pos	Neg	Neg	Pos	Not done	Neg	Pos
R.P.	Neg	Neg	Neg	Neg	Neg	Not done	Neg	Neg

[a]Neg = negative result; pos = positive result.

From Kirkpatrick, C. H., Rich, R. R., and Smith, T. K., *J. Clin. Invest.*, 51, 2984, 1972. With permission.

lymphocytes of patients who had recovered from malignant melanoma, close relatives, or other individuals whose lymphocytes were cytotoxic to culture of melanoma cells. No clinical improvement was observed in these patients.

Between the two distinctly different success rates in treating different tumors, other preliminary reports indicate varying degrees of success in treating tumor patients with TF_d.[42-44] A number of studies are now in progress on the value of TF_d in the treatment of cancer. It is perhaps too early to evaluate this form of immunotherapy because of the large number of variables: the type of tumor, the level of tumor burden, the immunologic competence of the patient, the general physical condition of the patient, the level of donor hypersensitivity, the quantity of TF_d used in treating the patient, and the duration of treatment.

A variety of congenital immunodeficiency diseases have also been treated with TF_d: Wiskott-Aldrich syndrome (WAS),[13,45,47] Swiss-type agammaglobulinemia,[48] ataxiatelangiectasia[13] (AT), X-linked immunodeficiency,[49] and others. Results have been variable.

Spitler et al.[47] reported that 14 of the 28 patients having WAS responded to TF_d therapy. They showed clinical improvement and developed skin reactivity and MIF responsiveness to antigens to which the TF_d lymphocyte donor was positive. The authors suggested that there may be two forms of WAS. Patients who respond to TF_d have below normal levels of blood monocytes with receptors for IgG. Even those patients responding to TF_d still may retain some evidence of the immune defect. Lymphocytes from patients treated with TF_d can be stimulated to synthesize MIF in vitro. They may not, however, respond by LT. Transfer of specific hypersensitivity and the temporary reconstitution of the immune defect may be due to two or more distinct components of TF_d.

Griscelli et al.[13] treated four patients having WAS, five having AT, and two with variable hypogammaglobulinemia. Some patients did convert from negative to positive skin reactors after receiving TF_d from donors sensitive to PPD, candidin, or SK-SD. As other investigators have also noted, in some of these patients in vitro assays of CMI (LT) also became positive, but did not always correlate with conversion of skin reactivity. Some clinical improvement was noted in patients. Two of the four WAS patients improved. AT patients had no improvement in neurologic signs. Patients with combined immunodeficiency disease showed no clinical improvement or change in delayed hypersensitivity status. Griscelli questions the role of TF_d in transferring donor-specific hypersensitivity. He suggests that TF_d may augment an already existing hypersensitivity.

Other immune deficiency states may not respond at all to TF_d. Lawlor et al.[50] could not reconstitute DH or in vitro correlate activity in patients with Nezelof syndrome, nor was there any clinical improvement in these patients.

Patients with Hodgkins' disease have impaired CMI. Attempts to restore DH by transfusion of leukocytes from normal donors have not been successful. The inability to reconstitute Hodgkins' disease patients with peripheral blood donor lymphocytes indicates a defect of greater magnitude than the absence of a sensitive lymphocyte population. Khan et al.[51] injected TF_d into eight Hodgkin patients in remission, following chemotherapy or radiation therapy and chemotherapy. Seven patients, also in remission, did not receive TF_d and were used as controls. The patients received TF_d over varying periods of time. The minimum total TF_d dose given was obtained from 30×10^8 leukocytes. The maximum TF_d dose given was obtained from 49×10^8 leukocytes. The authors measured DH by skin testing patients to mumps, PPD, histoplasmin, dermatophytin, varidase, and DNCB. Transfer of DH was achieved in five patients. Two patients developed DH to antigens to which the TF_d donors had demonstrated no DH. It is possible that either the donors or recipients may have had a low level of DH to these antigens. Control patients receiving no TF_d also demonstrated some skin reactivity to the anergy panel of antigens. The frequency of response to antigen was greater in the patients receiving TF_d than in those receiving no immunotherapy. The authors also reported that there was a slight increase in the level of E-rosette lymphocytes over that found in control patients receiving no TF_d (46.3 to 37.8%). The difference, however, was not statistically significant. TF_d appears to reconstitute Hodgkins' disease patients temporarily.

TF_{nd} has not been used to treat immunodeficiencies in humans. Its effectiveness and the specificity of the transfer with this material are yet to be determined.

In Vitro Transfer in the Human System

The demonstration of the in vitro transfer of CMI to specific antigens with TF_d or TF_{nd} to lymphocytes in culture has been extremely difficult. The approach which many investigators have used is to attempt to demonstrate recruitment of nonsensitive peripheral blood leukocytes. Nonsensitive peripheral blood leukocytes are incubated with TF_d or TF_{nd}. Antigen, to which the TF_d or TF_{nd} leukocyte donor is sensitive, is then added to the cultures. Control cultures are incubated without antigen or with an unrelated antigen. The cultures are subsequently assayed for the production of lymphokines, which have been demonstrated to be in vitro correlates of CMI. The two assays most frequently used are LT and MIF production. There are three points concerning the advisability of using such assays in attempting to detect recruitment of lymphocytes by TF_d or TF_{nd} in vitro.

1. The assays were originally designed to detect existing host sensitivity (the number of lymphocytes sensitive to the antigen obtained from a donor may be far greater than the limited number of cells which can be recruited in vitro by TF; the assays may, therefore, not be sufficiently sensitive).
2. The types, numbers, and stages of development of the cells which are recruited in vivo by TF_d may not be available in peripheral blood for in vitro recruitment.
3. The in vitro conditions may not be an adequate environment for the expression of a population of lymphocytes recruited by TF_d.

In vitro transfer of sensitivity to a number of microbial antigens using the indirect MIF assay of Thor and Dray[52] was demonstrated by Paque et al.[53] TF_L was prepared from the blood leukocytes of donors sensitive to PPD, histoplasmin, or coccidiodin. The TF_L, when incubated with nonsensitive human leukocytes in the presence of specific antigen, caused the lymphocytes to produce MIF. Subsequently, these authors reported that when the TF_L was dialyzed against distilled water, the TF_{nd} was capable of transferring hypersensitivity to nonsensitive human lymphocytes in vitro. The dialysate which should have contained TF_d did not transfer specific hypersensitivity in vitro.[54] Varying concentrations of the TF_d were found to be ineffective. The TF_L, TF_{nd}, or TF_d used in the in vitro assay was obtained from approximately 2.5 $\times$ 10^7 donor leukocytes. The TF preparations were incubated with 24 $\times$ 10^6 nonsensitive recipient lymphocytes. The concentration of TF preparation used was limited by the toxicity of the preparation. Higher concentrations of TF_L, TF_{nd}, or TF_d caused killing of the recipient lymphocytes.

Using the LT assay, Kniskern and Baram[75] attempted to demonstrate in vitro specific recruitment of nonsensitive human lymphocytes with portions of the same pools of TF_L, TF_{nd}, and TF_d which Paque et al. had used. They could not demonstrate any increase in tritiated thymidine (^{3}H-T) incorporation by nonsensitive lymphocytes incubated with any of the TF preparations and specific antigen.

Ascher et al.[14] reported on the recruitment of nonsensitive human lymphocytes by TF_d (using the LT assay) in nonsensitive leukocyte cultures incubated with TF_d and specific antigen. TF_d was prepared from the leukocytes of donors sensitive to either PPD, SK-SD, coccidiodin, or diphtheria toxoid. Recruitment of nonsensitive lymphocytes was observed only when the specific antigen and TF_d were incubated with the nonsensitive lymphocytes. Although the nonsensitive lymphocytes used for the recruitment by TF_d were obtained from skin-test-negative donors, these lymphocytes were slightly stimulated by the antigens in vitro prior to the addition of TF_d (Table 3). There is, therefore, some question as to the specificity of the recruitment capacity of the TF_d preparation. However, the augmentation effect was quite definitive. The authors noted that this effect could not be accomplished in vitro using TF_d which had been prepared in the usual manner: dialysis of the leukocyte lysates against distilled water followed by lyophilization of the dialysate. The only preparations which were effective in their studies were those dialyzed against small volumes of minimal essential medium for 18 hr. They designated these preparations TF_{dm} to distinguish them from TF_d. TF_{dm} obtained from leukocytes containing 6 $\times$ 10^6 lymphocytes was sufficient to induce increased ^{3}H-T incorporation in 1.5 to 2.0 $\times$ 10^6 lymphocytes in the presence of specific antigen.

Ascher and Andron,[55] in reevaluating the recruitment of lymphocytes by TF_{dm} using the LT assays, have in a subsequent series of experiments presented evidence indicating that the data of

TABLE 3

Tuberculin Stimulation of Nonimmune Lymphocytes: Range of Responses to Constant Amounts of TF_{dm} and PPD‡

Immune donor	Nonimmune recipient	No addition	PPD 4 μg	TF_{dm}[a]	TF_{dm}[a] + PPD 4 μg	Ratio[b]
JOS	MA	834 ± 304[c]	7,624 ± 3,581	298 ± 110	26,886 ± 3,808	3.53
JC	MA	204 ± 91	1,111 ± 440	484 ± 191	3,601 ± 1,602	3.24
JOS	JB	268 ± 64	2,697 ± 559	411 ± 191	8,534 ± 683	3.17
JC	BA	276 ± 45	3,495 ± 1,504	411 ± 18	10,451 ± 3,008	2.99
JK	PB	138 ± 22	1,688 ± 485		4,593 ± 495	2.94
DK	JM	401 ± 22	4,688 ± 236	392 ± 33	13,328 ± 2,157	2.84
JS	AL	538 ± 969	8,645 ± 969		23,634 ± 3,597	2.73
WS	MA	1,379 ± 100	16,633 ± 5,143		42,850 ± 5,944	2.58
JC	AL	960 ± 315	11,612 ± 3,249		29,269 ± 2,618	2.52
JOS	PB	106 ± 1	4,054 ± 210		8,938 ± 2,045	2.21
JOS	PO	202 ± 71	7,099 ± 1,994		14,927 ± 4,329	2.10
HUC	MA	214 ± 33	3,041 ± 659		6,534 ± 1,256	2.09
JOS	PO	360 ± 68	4,901 ± 1,326		9,469 ± 2,021	1.93
JOS	BA	1,806 ± 354	16,834 ± 4,831	2407 ± 701	32,224 ± 2,652	1.92

[a] TF_{dm} prepared from 6×10^6 lymphocytes.
[b] Ratio of (TF_{dm} + PPD)/(PPD alone).
[c] Mean and SD of triplicate or quadruplicate tubes in dPM of [^{14}C]thymidine incorporated.

From Ascher, M. S., Schneider, W. J., Valentine, F. T., and Lawrence, H. S., *Proc. Natl. Acad. Sci. U.S.A.*, 71, 1178, 1974. With permission.

Ascher et al.[14] may have reflected nonspecific augmentation of ^{3}H-T incorporation. Hamblin and Dumond[56] have also reported that the augmentation of ^{3}H-T incorporation by nonsensitive lymphocytes in the presence of TF_d and antigen is nonspecific. It is possible that both a specific recruitment and a nonspecific augmentation can occur.

Pizza et al.[57] have reported that human lymphocyte cell lines could be recruited by TF_d. TF_d prepared from the recruited lymphocyte lines could subsequently recruit fresh nonsensitive lymphocytes. The transfer is specific. This interesting study awaits verification by other investigators. Corroboration of this study on the in vitro production, as well as assay of TF_d, would be of value in illustrating the mechanisms of the induction of the hypersensitive state, and these cells would also serve as a source of TF_d.

C. Nonhuman In Vivo Models

Other TF model systems reported are nonhuman primates, guinea pigs, rats, dogs, and rabbits. In human TF studies, peripheral white blood cells have been the primary source of TF_d and TF_{nd}. However, in other species, peritoneal exudate cells, lymph node cells, and alveolar washings from lungs as well as peripheral white blood cells have been used as sources of TF_d and TF_{nd}. There are no reports evaluating thymus or bone marrow cells as a source of TF.

The in vivo transfer of DH using TF_L was first reported in the guinea pig system.[3] DNCB sensitivity was transferred to nonsensitive guinea pigs. Transfer of sensitivity was verified by skin test. The skin test reactivity could be elicited, however, only if the recipient's skin was treated with a chemical depilatory reagent. Subsequent reports indicate that transfer can be accomplished in the guinea pig system, but only if large numbers of lymphocytes are used to prepare TF_d. Burger and Jeter[16] produced pools of TF_d from peritoneal exudate, lymph node, and alveolar washing cells. These were incubated for 4 hr at 37°C. The incubation fluid obtained from 1×10^9 leukocytes was then injected intraperitoneally into nonsensitive recipient animals. If one considers the amount of TF_d given to a human recipient, the dose which the guinea pigs received may be > 150 to 300 human unit equivalents. Approximately 40% of the transfers have been successful.[76]

Nonhuman primates have recently been used as models for the in vivo, as well as in vitro, study of TF_d and TF_{nd}. The nonhuman primate models offer many advantages in that the animals are phylogenetically closely related to humans. Responses in these animals may more closely approximate those obtained in humans than those obtained in other species. Because of the antigenic relatedness of many tissues, human autoimmune disease could be studied. TF_d or TF_{nd} prepared from the leukocytes of patients with cell-mediated autoimmune disease could be used in an attempt to transfer the disease to nonhuman primates. The size of some species of nonhuman primates also makes them good donors of large numbers of leukocytes essential for fractionation studies and in vitro studies of TF. However, DH skin reactions are difficult to obtain in nonhuman primates, and investigators must often depend on in vitro correlate assays of CMI.

In vivo transfer of OH to a number of antigens has been reported using both TF_d and TF_{nd}. Maddison et al.[26] transferred DH to rhesus monkeys in vivo. Both humans and rhesus monkeys served as leukocyte donors for the preparation of PPD-TF_{nd} and PPD-TF_d. Rhesus monkeys previously sensitized to *Schistosoma mansoni* (SM) were used as leukocyte donors for SM-TF_{nd} and SM-TF_d. The animals were skin tested with PPD and SM 48 hr after receiving either TF preparation, and the sites were biopsied 27 hr later. The biopsies were sectioned, and the stained preparations were examined for cellular infiltration. Based on the degree of infiltration, a site was classified as positive or negative for DH. Rhesus monkeys receiving human or rhesus monkey PPD-TF_{nd} or PPD-TF_d developed positive skin tests to PPD, but did not develop DH reactions to SM. Conversely, monkeys injected with SM-TF_d or SM-TF_{nd} developed positive skin tests to SM antigen but not to PPD. This would argue for antigen specificity of the TF preparations. However, one of four animals receiving TF_d prepared from nonsensitized moneky leukocytes and one of two monkeys receiving TF_{nd} prepared from nonsensitized monkeys developed positive skin test to PPD. This lack of specificity may reflect exisiting recipient sensitivity to antigens elevated by the adjuvant effect of TF. Maddison et al. also observed that lymphocytes from the PPD recipients could be stimulated in vitro 3 weeks later with PPD. It is important to consider that these lymphocytes were obtained from animals that had not only received PPD-TF_{nd} or PPD-TF_d, but had been skin tested with PPD which may have sensitized the recipients.

Baram and Condoulis[73] could demonstrate no LT with PPD or KLH in lymphocytes from ten rhesus monkeys which had received KLH, PPD-TF_d or lymphocytes from three rhesus monkeys which had received KLH, PPD-TF_{nd}. They did, however, find that lymphocytes from TF_d recipients and TF_{nd} recipients produced MIF in culture when stimulated with specific antigens. Their failure to demonstrate LT may be due to the quantity of TF_d and TF_{nd} injected into the recipients.

Maddison et al.[26] reported that human TF was more effective than rhesus monkey TF in transferring DH to recipient animals. This may indicate that the reason for the more readily demonstrable transfers in the human system may be due to the quantity or stability of the TF in the preparations.

It is important to note that the histological examination of skin biopsy material may have many disadvantages in assessing DH. The skin testing agent may induce a chronic inflammatory response in the tissue indistinguishable from a DH reaction. Needle tracks due to the injection or tissue abrasion in preparing the skin site for injection may also result in an infiltrate not readily distinguished from DH.

The human-to-rhesus transfer in vivo using TF_d has been confirmed by Gallin and Kirkpatrick.[11] They were able to transfer DH to three antigens: mumps, candida, and SK-SD. TF_d was prepared from hypersensitive human leukocytes. The human donors were only moderately sensitive by skin test to the antigens. The TF preparation also contained substances chemotactic for mononuclear cells and neutrophils in vivo and for granulocytes in vitro. Fractionation of this human TF_d resulted in a fraction having transfer capacity for specific antigen as well as chemotactic properties.

Steele et al.,[58] using TF_d prepared from either human or baboons, reported on the transfer of DH to baboons, cebus monkeys, or marmosets. Sensitivity to candida, mumps, herpes virus type 1, tetanus toxoid, and coccidiodin were transferred, although not consistently. Recipients received the TF_d from 1×10^8 cells/7K of body weight. Human TF_d was more effective in transferring hypersensitivity than was the baboon-derived TF_d. In these experiments, negative TF_d was not used to inject control groups of animals. The authors did not feel that there was any difficulty in using the skin test as an assay for DH.

Baram and Condoulis[73] have transferred hypersensitivity to KLH in 50% of rhesus monkey recipients (MIF assay) using rhesus monkey TF_d

or TF_{nd}. They also detected immediate hypersensitivity responses in the recipients up to 168 days after the first injection of TF.[59] No DH skin tests were positive.

A model system using the rat was devised by Liburd et al.[17] The authors prepared TF_d and TF_{nd} from the pooled peripheral white blood cells, spleen, and lymph nodes obtained from rats sensitive to *Eimerio niescholzi,* the parasite causing rat coccidiosis. Six rats each received i.p. injections of 200 mg of TF_d which was obtained from approximately 20×10^8 leukocytes. The animals were then challenged with 1,000 sporulated oocysts, and oocysts were counted in the stool for 7 to 12 days. The animals were rechallenged 15 days later with 50,000 oocysts and the number of oocyst counted daily for 15 days. Animals receiving specific TF_d had a decrease in the number of oocysts compared to animals receiving TF_{nd}. These TF_d recipients had no oocysts in the stool upon rechallenge with 50,000 oocysts. These studies are interesting. Elaboration of the system is necessary before these results can be evaluated. It does increase the opportunity for using in vivo assays other than skin testing.

A number of studies have recently been undertaken to investigate the human-to-guinea pig transfer system. Investigators are not in agreement on the capacity of human TF_d to transfer donor sensitivity to recipient guinea pigs. It is difficult to determine if the observed skin test reactions in guinea pig recipients of human TF_d reflect specific transfer, augmentation, or nonspecific responses.[60,61]

Nonhuman In Vitro Models

Baram and Condoulis[62] first reported the transfer of KLH CMI from sensitized rhesus monkeys, using peripheral leukocyte TF_{nd}, to nonsensitive rhesus monkey lymphocytes in vitro. When either KLH-TF_L or KLH-TF_{nd} was incubated with nonsensitive rhesus monkey lymphocytes in culture, the addition of KLH resulted in increased LT. However, KLH-TF_d was not effective in recruiting lymphocytes in the presence of KLH using the LT assay. Curiously, although KLH-TF_L and KLH-TF_{nd} were effective in in vitro recruitment, PPD-TF_L and PPD-TF_{nd} were not capable of in vitro recruitment in the presence of PPD. The unique phenomenon observed with KLH-TF_L and KLH-TF_{nd} may be related to the high molecular weight of KLH. The reactions were specific; only KLH could stimulate the lymphocytes incubated with KLH-TF_L or KLH-TF_{nd}. Human blood lymphocytes were also specifically recruited using rhesus monkey KLH-TF_L or KLH-TF_{nd} but were not recruited by rhesus KLH-TF_d or PPD-$TF_{L,nd,d}$.

Transfer between more distantly related species has also been reported. Adler et al.[63] prepared TF_d from the blood leukocytes of a human donor sensitive to PPD. This PPD-TF_d, incubated with murine spleen cells in the presence of PPD, stimulated LT. In the absence of PPD, no lymphocyte stimulation was observed. Unfortunately, only one PPD-TF_d preparation was used. TF_d from nonsensitive human donors was used as a control in this experiment. The assessment of transfer capacity of only one positive PPD-TF_d preparation makes it difficult to evaluate the experiment. A subsequent report from the same laboratory[64] in which the human-to-mouse transfer system was reinvestigated questions the specificity of the transfer. PPD-TF_d, PPD-TF_{nd}, and TF_L prepared from the peripheral blood of patients with pulmonary tuberculosis and from donors negative to intermediate strength PPD (N-TF_d and N-TF_{nd}) were evaluated for the in vitro transfer of tuberculin hypersensitivity to murine spleen cells. The authors indicated that PPD-TF_L and PPD-TF_d appeared to augment PPD stimulation of murine spleen cells. The cells reacted by increased incorporation of tritiated thymidine in the presence of PPD-TF_d and PPD. However, the reaction was not affected by treatment with anti-θ serum and complement. This would imply that the PPD-TF_d may act on a subpopulation of B cells in the augmentation of LT.

Baram and Condoulis[59] attempted to recruit nonsensitive rhesus monkey lymphocytes in vitro with rhesus monkey KLH-TF_d. They could not stimulate LT in the presence of KLH-TF_d and KLH. The MIF assay was also negative. However, when KLH-TF_d was fractionated on Sephadex G-25, one of the fractions was found to recruit lymphocytes in vitro to produce MIF in the presence of KLH.

The reports on the demonstration of the in vitro recruitment of nonsensitive lymphocytes by TF_d or TF_{nd} are not satisfactory. At best LT assays indicate that the effect of TF preparations is nonspecific. The MIF assay does indicate some specificity in the TF recruitment of lymphocytes in vitro. However, the assay is difficult to run and the results are not uniformly reliable. Salaman,[65] using human PPD-TF_d to recruit guinea pig peritoneal exudate lymphocytes, found that the increased migration or inhibition of migration of macrophages was dependent on the degree of DH of the TF_d donor and the quantity of PPD used in the assay. The dilemma remains. Although TF preparations cause recipients to develop DH, there is still the question of the specificity of transfer, and this question is reinforced by the failure to consistently demonstrate in vitro recruitment of lymphocytes by TF_d and TF_{nd}.

ISOLATION AND IDENTIFICATION OF TF_{nd} AND TF_d

It is, of course, important to determine the composition of both TF_{nd} and TF_d. Fractionation of TF preparations and isolation of components having distinct properties could permit their quantitation. Their clinical use and value could then be better monitored and evaluated. The relationship of TF_{nd} and TF_d to each other could be determined. Finally, purified preparation of TF would be of inestimable value in studying T-cell recruitment and the mechanism by which T cells recognize antigen.

Progress in the purification of TF_{nd} and TF_d has been exceedingly slow. This has been due to the large number of recipients needed for the in vivo assay of samples prepared in the fractionation studies and the difficulty in establishing specific reliable in vitro assays.

Fractionation of TF_{nd} and TF_d

The first attempts to fractionate TF_L were made by Lawrence.[65] He demonstrated that dialysates of lysed human leukocytes contained TF activity. He was unable to detect any transfer factor activity in the nondialyzable portion of the lysate. He further separated the dialyzable portion into two fractions using molecular sieving chromatography. Chromatographed on Sephadex G-25, the TF_d activity was found in the second of two peaks, indicating that it was a low molecular weight substance. Baram et al.[7,8] fractionated human PPD-TF_L on Sephadex G-200 and found transfer factor activity in two peaks (Figure 2). The A′ component contained TF_{nd} and the C component contained the TF_d. The C component was further chromatographed on Sephadex G-25,

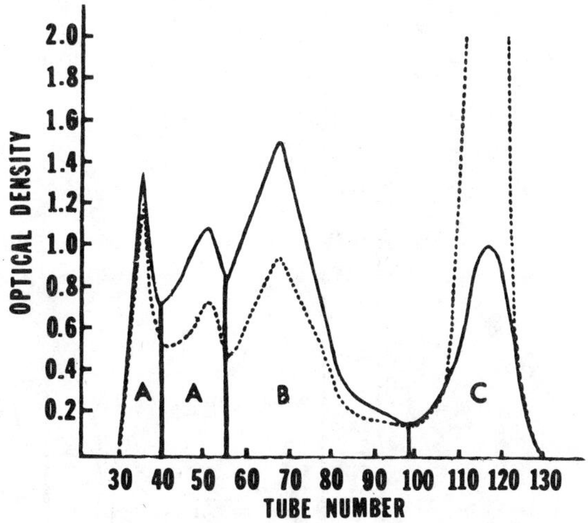

FIGURE 2. Sephadex G-200 chromatography of human PPD-TF_L. The column dimensions were 100 × 2.5 cm. The A′ and C peaks transferred PPD hypersensitivity to human recipients. Solid lines indicate absorbance at 280 nm; dotted lines indicate absorbance at 260 nm. (From Baram, P., Yuan, L., and Mosko, M. M., *J. Immunol.*, 97, 407, 1966. With permission.)

and the C′-5 peak (Figure 3) was found to transfer DH to PPD in PPD negative human recipients.

A number of investigators have also reported on the fractionation of human and nonhuman primate TF_d using either Sephadex G-25 or G-10 or Bio-Gel.[10,11,31,59,67-69] Although there are slight differences in the chromatograms obtained by various investigators, these are in part due to the eluent used. Ammonium bicarbonate and phosphate buffers have been used most frequently, but adequate separation is also obtained when distilled water is used as the eluent. A number of investigators are in agreement that the peaks which transfer DH have an unusually high 260:280 absorbance ratio which may in part be due to substances other than polyribonucleotides.

Composition of TF_{nd} and TF_d Fractions

TF_{nd}

TF_{nd} contains immunoglobulins and other proteins. It also contains large quantities of degraded RNA, approximately 4S. No intact RNA is detectable. This is to be expected. Lazda and Baram[70] extracted the RNA from rhesus monkey KLH-TF_{nd}. They evaluated the ability of this RNA to recruit lymphocytes in vitro in the presence of KLH. Although elevation of tritiated

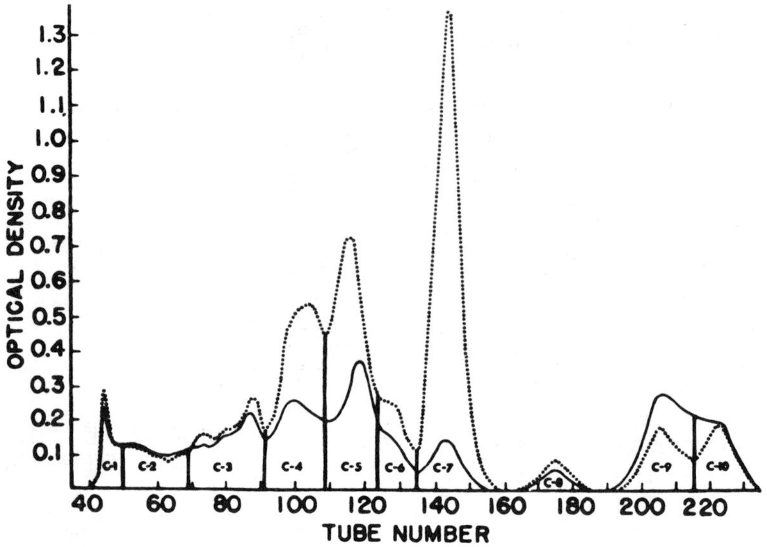

FIGURE 3. Sephadex G-25 chromatography of human PPD-TF_d. The C-5 peaks transferred PPD hypersensitivity to human recipients. Solid lines indicate absorbance at 280 nm; dotted lines indicate absorbance at 260 nm. (From Baram, P., Yuan, L., and Mosko, M. M., *J. Immunol.*, 97, 407, 1966. With permission.)

thymidine incorporation was observed, the RNA was not as effective as unfractionated TF_{nd}.

Treatment of rhesus monkey KLH-TF_{nd} with pronase (for 22 hr at 37° C) caused hydrolysis of the proteins with a concomitant loss of recruitment capacity. However, portions of TF_{nd} incubated at the same temperature without pronase also lost recruitment capacity. It is, therefore, not possible at this time to ascribe TF_{nd} activity either to the protein or RNA. No antigen has been detected in TF_{nd} preparations. Because of the lack of purity, estimates of molecular weight have little significance. TF_{nd} contains proteins and RNA with molecular weights ranging from 70,000 to 300,000.

TF_d

TF_d, regardless of the species source, contains polynucleotides, polypeptides of varying molecular weights, and inorganic salts. Although TF_d preparations are not degraded by ribonuclease or proteolytic enzymes, one cannot rule out the likelihood that TF_d may be a partially hydrolyzed fragment resistant to further enzyme degradation. Spitler[71] reported that TF_d treated with pronase inactivated the transfer capacity of the preparation. However, ribonuclease may have contaminated the pronase, and it is important to establish if one or both enzymes were involved in the inactivation. As with TF_{nd}, there is uncertainty of the molecular weight. Molecular weight determinations varying from 2,000 to < 10,000 have been reported. These have usually been made on the bases of Sephadex chromatography and ultracentrifugation studies. Since the composition of TF_d is also unknown, one can only ascertain at the present time that TF_d has a low molecular weight.

There is no evidence that TF_d contains antigen. This, however, does not rule out the possibility that very small undetectable quantities of antigen are present in the TF_d preparations, which may be sufficient to sensitize the recipient. Further, methods normally employed to detect antigen may be inadequate since the antigen may be sequestered by polynucleotide or polypeptide complexes.

TF_d does not appear to be immunogenic. This is based primarily on the failure of repeated injections of human TF_d into human recipients to cause any observable allergic reaction. TF_d emulsified with complete Freunds' adjuvant was found to cause the production of two antibodies in rabbits, detectable by immunoelectrophoresis.[7] It is not known if the antibodies are directed against that moiety which transfers donor-specific hypersensitivity, but they may well be of value in further purification studies of TF_d.

DISCUSSION

There is evidence that two fractions derived from leukocyte lysates have the capacity to transfer donor-specific hypersensitivity and, under some conditions, temporarily restore immunocompetence to immunologically compromised individuals. Although evidence of the specificity of the transfer with TF_{nd} and TF_d continues to accumulate, it is not yet definitive. This may, in part, be due to the melange of substances present in the two preparations. Both histologic spleen changes and altered in vitro responses to mitogens have been observed in mice receiving either human or mouse TF_d.[9] Changes in the number of T cells in the peripheral circulation and changes in MLR[72] have also been observed in human recipients of TF_d. Increased numbers of B cells have also been reported.[73] These events may be entirely unrelated to that phenomenon which results in the transfer of donor hypersensitivity, but may be of consequence in therapeutic use. Polynucleotides and degraded RNA may exert an adjuvant-like effect which results in temporary reconstitution of the immune responsiveness of the compromised recipient. It may also be the reason for the observed occasional conversion of skin test reactivity to an antigen to which neither the donor nor recipient was overtly sensitive prior to the transfer.

Clinically, regardless of the specificity of the transfer, TF_d has been shown to have potential as an immunotherapeutic agent with few side effects. It may eventually even have value as a diagnostic tool, helping to define immunologic diseases.

Reports on the value of TF_d as an immunotherapeutic agent are on occasion contradictory, but this may be due partly to the poorly defined unit of TF_d, the variety of donors used as a leukocyte source, the compositional differences between lots of TF_d, and, finally, the enthusiasm of the clinician immunologist. Fortunately, more double blind studies are being developed to control the latter variable.

Although TF has been with us for almost two decades, there is still no clear idea of the composition of TF_{nd} and TF_d and their relationship to each other. Are they distinct, naturally occurring entities within the intact cell or are they the artifacts of cell disruption, which still retain biologic activity? Perhaps they fortuitously gain biologic activity as degraded substances where none existed in the intact cell. If TF_{nd} is degraded RNA and TF_d is an oligonucleotide, they may both be degradation products of (immunogenic) RNA. This would make TF_d the more valuable, if of equal potence with immunogenic RNA. Low molecular weight TF_d carries little chance of transferring portions of viral genomes.

If the TF_{nd} and TF_d are protein or polypeptide, then it is important to determine if antigen is present. If antigen is not present, then these moieties are a key to as yet unexplored concepts in the cell-mediated immune phenomenon. It is, of course, premature to develop complex hypotheses about the mechanism of action of TF_{nd} and TF_d. It has variously been suggested that TF_{nd} and TF_d may act as lymphocyte membrane receptors, lymphocyte derepressors, highly processed antigens, portions of messenger RNA, and variable regions of antibody molecules. It is difficult at this time to give emphasis to any one hypothesis. Research on TF has increased rapidly within the last few years, and it may soon be possible to address many of the problems raised by the existence of TF_{nd} and TF_d.

REFERENCES

1. **Landsteiner, K. and Chase, M. W.,** Experiments on transfer of cutaneous sensitivity to sample compounds, *Proc. Soc. Exp. Biol. Med.,* 49, 688, 1942.
2. **Chase, M. W.,** The cellular transfer of cutaneous hypersensitivity to tuberculin, *Proc. Soc. Exp. Biol. Med.,* 59, 134, 1945.
3. **Jeter, W. S., Tremaine, M. M., and Seebohm, P. M.,** Passive transfer of delayed hypersensitivity to 2,4-dinitrochlorobenzene in guinea pigs with leukocytic extracts, *Proc. Soc. Exp. Biol. Med.,* 86, 251, 1954.
4. **Lawrence, H. S.,** The cellular transfer of cutaneous hypersensitivity to tuberculin in man, *Proc. Soc. Exp. Biol. Med.,* 71, 516, 1949.
5. **Lawrence, H. S.,** The transfer in humans of delayed skin sensitivity to streptococcal M substance and to tuberculin with disrupted leucocytes, *J. Clin. Invest.,* 34, 219, 1955.
6. **Lawrence, H. S., Davis, J., Franklin, E. C., and Zweiman, B.,** Transfer of immunological information in humans with dialysates of leukocyte extracts, *Trans. Assoc. Am. Physicians,* 76, 84, 1963.
7. **Baram, P. and Mosko, M. M.,** A dialysable fraction from tuberculin sensitive human white blood cells capable of inducing tuberculin-delayed hypersensitivity in negative recipients, *Immunology,* 8, 461, 1965.
8. **Baram, P., Yuan, L., and Mosko, M. M.,** Studies on the transfer of human delayed-type hypersensitivity. I. Partial purification and characterization of two active components, *J. Immunol.,* 97, 407, 1966.
9. **Huard, T. K. and Baram, P.,** Non-specific effects of human dialyzable transfer factor, in *Transfer Factor: Basic Properties and Clincial Application,* Ascher, M. S., Kirkpatrick, C. H., and Gottlieb, A. A., Eds., Academic Press, New York, 1976.
10. **Burger, D. R., Nolte, J. E., Vandenbark, A. A., and Vetto, K. M.,** Biological activity of subfractions of human transfer factor, in *Transfer Factor: Basic Properties and Clinical Application,* Ascher, M. S., Kirkpatrick, C. H., and Gottlieb, A. A., Eds., Academic Press, New York, 1976.
11. **Gallin, J. I. and Kirkpatrick, C. H.,** Chemotactic activity in dialyzable transfer factor, *Proc. Natl. Acad. Sci. U.S.A.,* 71, 498, 1974.
12. **Wybran, J., Levin, A. S., Spitler, L. E., and Fudenberg, H. H.,** Rosette-forming cells, immunologic deficiency diseases and transfer factor, *N. Engl. J. Med.,* 288, 710, 1973.
13. **Griscelli, C.,** Transfer factor therapy in immunodeficiencies, in *Clinical Immunobiology,* Vol. 2, Bach, F. H. and Good, R. A., Eds., Academic Press, New York, 1974.
14. **Ascher, M. S., Schneider, W. J., Valentine, F. T., and Lawrence, H. S.,** *In vitro* properties of leucocyte dialysates containing transfer factor, *Proc. Natl. Acad. Sci. U.S.A.,* 71, 1178, 1974.
15. **Rocklin, R. E., MacDermott, R. P., Chess, L., Schlossman, S. F., and David, J. R.,** Studies on mediator production by highly purified human T and B lymphocytes, *J. Exp. Med.,* 140, 1303, 1974.
16. **Burger, D. R. and Jeter, W. S.,** Cell-free passive transfer of delayed hypersensitivity to chemicals in guinea pigs, *Infect. Immun.,* 4, 575, 1971.
17. **Liburd, E. M., Pabst, H. F., and Armstrong, W. D.,** Transfer factor in rat coccuiliosis, *Cell. Immunol.,* 5, 487, 1972.
18. **Kirkpatrick, C. H., Rich, R. R., and Smith, T. K.,** Effect of transfer factor on lymphocytic function in anergic patients, *J. Clin. Invest.,* 51, 2984, 1972.
19. **Spitler, L. E., Levin, A. S., and Fudenberg, H. H.,** Transfer factor, in *Clinical Immunology,* Vol. 2, Bach, F. H. and Good, R. A., Eds., Academic Press, New York, 1974.
20. **Lawrence, H. S. and Pappenheimer, A. M.,** Transfer of delayed hypersensitivity to diphtheria toxin in man, *J. Exp. Med.,* 104, 321, 1956.
21. **Lawrence, H. S. and Pappenheimer, A. M.,** Effect of specific antigen on release from human leukocytes of the factor concerned in transfer of delayed hypersensitivity, *J. Clin. Invest.,* 36, 908, 1957.
22. **Graybill, J. R., Silva, J., Jr., Alford, R. H., and Thor, D. E.,** Immunologic and clinical improvement of progressive coccidiodomycosis following administration of transfer factor, *Cell. Immunol.,* 8, 120, 1973.
23. **Oliveira-Lima, O.,** Passive transfer of the delayed dermal sensitivity to tuberculin by means of blood leukocytes. The impedient effect of specific desensitization, *Am. Rev. Tuberc.,* 78, 346, 1958.
24. **Spitler, L. E.,** personal communication.
25. **Burger, D. R., Vetto, R. M., and Malley, A.,** Transfer factor from guinea pigs sensitive to dinitrochlorobenzene – absence of superantigen properties, *Science,* 175, 1473, 1972.
26. **Maddison, S. E., Hicklin, M. D., Conway, B. P., and Kagan, I. G.,** Transfer factor-delayed hypersensitivity to *Schistosoma mansoni* and tuberculin in *Macaca mulatta, Science,* 178, 757, 1972.
27. **Rocklin, R. E., Chilgren, R. A., Hong, R., and David, J. R.,** The transfer of cellular hypersensitivity in chronic mucocutaneous candidiasis monitored *in vivo* and *in vitro, Cell. Immunol.,* 1, 290, 1970.
28. **Feigen, R. D., Shackelford, P. G., Eisen, S., Spitler, L. E., Pickering, L. K., and Anderson, D. C.,** Treatment of mucocutaneous candidiasis with transfer factor, *Pediatrics,* 53, 63, 1974.
29. **Lawrence, H. S.,** Transfer factor in cellular immunity, The Harvey Lectures, Series 68, 261, 1974.
30. **Maurer, P. H.,** Immunologic studies with ethylene oxide-treated human serum, *J. Exp. Med.,* 113, 1029, 1961.
31. **Zukerman, K. S., Neidhart, J. A., Balcerzak, S. F., and LoBuglio, A. F.,** Immunologic specificity of transfer factor, *J. Clin. Invest.,* 54, 997, 1974.

32. **Brandriss, M. W.,** Attempt to transfer contact hypersensitivity in man with dialysate of peripheral leucocytes, *J. Clin. Invest.,* 47, 2152, 1968.
33. **Arala-Chaves, M. P. and Pinto, A. S.,** Transfer factor with dinitrochorobenzene. Several remarks, *Int. Arch. Allergy Appl. Immunol.,* 43, 410, 1972.
34. **Goldberg, R. S., Bluestone, R., Barnett, E. V., and Landau, J. W.,** Studies on lymphocyte and monocyte function in chronic mucocutaneous candidiasis, *Clin. Exp. Immunol.,* 8, 37, 1971.
35. **Catenzaro, A. and Spitler, L. E.,** Cell mediated immunity in lung disease, in *Immunologic and Infectious Reactions in the Lung,* Kirkpatrick, C. H. and Reynolds, H. Y., Eds., Marcel Dekke, New York, 1976.
36. **Valdimarsson, H., Wood, C. B. S., Hobbs, J. R., and Holt, P. J. C.,** Immunological features in a case of chronic granulomatous candidiasis and its treatment with transfer factor, *Clin. Exp. Immunol.,* 11, 151, 1972.
37. **Pabst, H. F. and Swanson, R.,** Successful treatment of candidiasis with transfer factor, *Br. Med. J.,* 2, 442, 1972.
38. **Bullock, W. E., Fields, J. P., and Brandriss, M. W.,** An evaluation of transfer factor as immunotherapy for patients with lepromatous leprosy, *N. Engl. J. Med.,* 287, 1053, 1972.
39. **Hastings, R. C., Morales, N. J., Shannon, E. J., and Jacobson, R. R.,** Transfer factor in leprosy, in *Transfer Factor: Basic Properties and Clinical Application,* Ascher, M. S., Kirkpatrick, C. H., and Gottlieb, A. A., Eds., Academic Press, New York, 1976.
40. **Levin, A. S., Byers, V. S., Fudenberg, H. A., Wybran, J., Hackett, A. J., Johnston, J. O., and Spitler, L. E.,** Immunologic parameters before and during immunotherapy with tumor specific transfer factor, *J. Clin. Invest.,* 55, 487, 1975.
41. **Price, F. B., Hewlett, J. S., Deodhar, S. D., and Barna, B.,** The therapy of malignant melanoma with transfer factor. A preliminary report, *Cleve. Clin. Q.,* 41, 1, 1974.
42. **Oettgen, H., Old, L., Farrow, J., Valentine, F., Lawrence, H. S., and Thomas, L.,** Effects of transfer factor in cancer patients, *J. Clin. Invest.,* 50, 71, 1971.
43. **LoBuglio, A. F., Neidhart, J. A., Hilberg, R. W., Metz, E. N., and Balcerzak, S. P.,** The effect of transfer factor therapy on tumor immunity in alveolar soft part sarcoma, *Cell. Immunol.,* 7, 159, 1973.
44. **Spitler, L. E., Wybran, J., Fudenberg, H. H., Levin, A. S., Lewis, M., and Horn, L.,**Transfer factor therapy of malignant melanoma, *Clin. Res.,* 21, 653, 1973.
45. **Levin, A. S., Spitler, L. E., Stites, D. P., and Fudenberg, H. H.,** Wiskott-Aldrich Syndrome, a genetically determined cellular immunologic deficiency – clinical and laboratory responses to therapy with transfer factor, *Proc. Natl. Acad. Sci. U.S.A.,* 67, 821, 1970.
46. **Spitler, L. E., Levin, A. S., Stites, D. P., Fudenberg, H. H., Pirofsky, B., August, C. S., Stiehm, E. R., Hitzig, W. H., and Gatti, R. A.,** The Wiskott-Aldrich Syndrome. Results of transfer factor therapy, *J. Clin. Invest.,* 51, 3216, 1972.
47. **Spitler, L. E., Levin, A. S., and Fudenberg, H. H.,** Transfer factor, in *Clinical Immunobiology,* Bach, F. H. and Good, R. A., Eds., Academic Press, New York, 1974.
48. **Hitzig, W. H., Fontanellay, H. P., Muntener, U., Paul, S., Spitler, L. E., and Fudenberg, H. H.,** Immunologische Grundlagen und theropeutische Erfahrungen, *Schweiz. Med. Wochenschr.,* 102, 1237, 1972.
49. **Goldblum, R. M., Lord, R. A., Dupress, E., Weinberg, A. G., and Goldman, A. S.,** Transfer factor induced delayed hypersensitivity in X-linked combined immunodeficiency, *Cell. Immunol.,* 9, 297, 1973.
50. **Lawlor, G. H., Ammann, A. J., Wright, W. C., LaFianchi, S. F., Galstrom, D., and Stiehm, E. R.,** Clinical and immunologic variability in cellular immunodeficiency with immunoglobulins, *Clin. Res.,* 21, 306, 1973.
51. **Khan, A., Hill, J. M., MacLellan, A., Loeb, E., Hill, N. O., and Thaxton, S.,** Improvement in delayed hypersensitivity in Hodgkins' disease with transfer factor: lymphapheresis and cellular immune reactions of normal donors, *Cancer,* 36, 86, 1975.
52. **Thor, D. E. and Dray, S.,** The cell-migration-inhibition correlate of delayed hypersensitivity. Conversion of human nonsensitive lymph node cells to sensitive cells with an RNA extract, *J. Immunol.,* 101, 469, 1968.
53. **Paque, R. E., Kniskern, P. J., Dray, S., and Baram, P.,** *In vitro* studies with "transfer factor": transfer of the cell migration inhibition correlate of delayed hypersensitivity in humans with cell lysates from humans sensitized to histoplasmin, coccidiodin, or PPD, *J. Immunol.,* 103, 1014, 1969.
54. **Paque, R. E., Dray, S., Kniskern, P., and Baram, P., II,** Transfer of the cell migration inhibition correlate of delayed hypersensitivity in human with nondialyzable components of leucocyte lysates from humans sensitized to histoplasmin and coccidiodin, *Cell. Immunol.,* 6, 368, 1973.
55. **Ascher, M. S. and Andron, L. A.,** *In vitro* properties of leucocyte dialysates containing transfer factor: micro method and recent findings, in *Transfer Factor: Basic Properties and Clinical Application,* Asher, M. S., Kirkpatrick, C. H., and Gottlieb, A. A., Eds., Academic Press, New York, 1976.
56. **Hamblin, A., Maini, R. N., and Dumonde, D. E.,** Antigen-dependent augmentation of lymphocyte transformation by dialysable transfer factor, in *Transfer Factor: Basic Properties and Clinical Application,* Ascher, M. S., Kirkpatric, C. H., and Gottlieb, A. A., Eds., Academic Press, New York, 1976.
57. **Pizza, G., Viza, D., Boucheix, C., and Corrado, F.,** Studies with *in vitro* produced transfer factor, in *Transfer Factor: Basic Properties and Clincial Application,* Ascher M. S., Kirkpatrick, C. H., and Gottlieb, A. A., Eds., Academic Press, New York, 1976.
58. **Steele, R. W., Eichberg, J. W., Heberling, R. L., Eller, J. J., Kalter, S. S., and Kniker, W. T.,** Transfer of cellular reactivity to 3 nonhuman primate species with human and baboon transfer factor, *Cell. Immunol.,* in press.

59. **Baram, P. and Condoulis, W.,** Studies on rhesus monkey nondialyzable and dialyzable transfer factors, *Transplant. Proc.,* 6, 209, 1974.
60. **Vandenbark, A. A., Burger, D. R., and Vetto, R. M.,** Human transfer factor: trials with an assay in guinea pigs, in *Transfer Factor: Basic Properties and Clinical Application,* Ascher, M. S., Kirkpatrick, C. H., and Gottlieb, A. A., Eds., Academic Press, New York, 1976.
61. **Welch, T. M., Wilson, G. B., and Fudenberg, H. H.,** Human transfer factor in guinea pigs: further studies, in *Transfer Factor: Basic Properties and Clinical Application,* Ascher, M. S., Kirkpatrick, C. H., and Gottlieb, A. A., Eds., Academic Press, New York, 1976.
62. **Baram, P. and Condoulis, W.,** The *in vitro* transfer of delayed hypersensitivity to rhesus monkey and human leukocytes with transfer factor obtained from rhesus monkey peripheral white blood cells, *J. Immunol.,* 104, 769, 1970.
63. **Adler, W. H., Takigushi, T., Marsh, B., and Smith, R. T.,** Cellular recognition of mouse lymphocytes *in vitro.* I. Definition of a new technique and results of stimulation by phytohemagglutinin and specific antigens, *J. Exp. Med.,* 131, 1049, 1970.
64. **Palmer, D. W. and Smith, R. T.,** Augmentation of PPD- and LPS-induced T-independent DNA synthesis in normal mouse spleen cells by leucocyte-lysates from tuberculosis patients, *Cell. Immunol.,* 13, 196, 1974.
65. **Salaman, M. R.,** Studies on the transfer factor of delayed hypersensitivity. Effect of dialysable leucocyte extracts from people of known tuberculin sensitivity on the migration of normal guinea-pig macrophages in the presence of antigen, *Immunology,* 26, 1069, 1974.
66. **Lawrence, H. S., Al-Askari, S., David, J., Franklin, E. C., and Zweiman, B.,** Transfer of immunological information in humans with dialysates of leucocyte extracts, *Trans. Assoc. Am. Physicians,* 76, 84, 1963.
67, **Arala-Chaves, M. P., Lebacq, E. G., and Heremans, J. F.,** Fractionation of human leucocyte extracts transferring delayed hypersensitivity to tuberculin, *Int. Arch. Allergy,* 31, 353, 1967.
68. **Gottlieb, A. A., Foster, L. F., Sarto, K., Sutcliffe, F., Oliver, T., Cullen, M., Wrigley, P., Fairley, G. H., Lopez, M., and Salvaggio, J.,** Partial purification of components of human transfer factor, in *Transfer Factor: Basic Properties and Clinical Application,* Ascher, M. S., Kirkpatrick, C. H., and Gottlieb, A. A., Eds., Academic Press, New York, 1976.
69. **Reymond, J.-F. and Grob, P. J.,** Transfer factor preparations: some physico-chemical and biological properties, *Vox Sang.,* 29, 338, 1975.
70. **Lazda, K. and Baram, P.,** unpublished data.
71. **Spitler, L. E.,** *Work on Basic Properties and Clinical Applications of the Transfer Factor,* Kirkpatrick, C. H. and Refkind, D., Eds., Academic Press, New York, 1973, 99.
72. **DuPont, B., Ballow, M., Hansen, J. A., Quick, C., Yunis, E. J., and Good, R. A.,** Effect of transfer factor therapy on mixed lymphocyte culture reactivity, *Proc. Natl. Acad. Sci. U.S.A.,* 71, 867, 1974.
73. **Condoulis, W. and Baram, P.,** unpublished results.
74. **Gelfand, E. W., Baumal, R., Huber, J., Crookston, M. C., Shumak, K. H.,** Gammopathy and lymphoproliferation after transfer factor in severe combined immunodeficiency disease, *N. Engl. J. Med.,* 289, 1385, 1973.
75. **Kniskern, P. J. and Baram, P.,** unpublished data.
76. **Burger, D. R.,** personal communication.

THE AUGMENTATION OF TUMOR-SPECIFIC IMMUNITY IN VITRO AND IN VIVO WITH IMMUNE RNA*

Peter J. Deckers, Bosco S. Wang, and John A. Mannick

It has been conclusively demonstrated that the lymphoid cells of the spleen, regional lymph nodes, thoracic duct lymph, and peritoneal cavity of animals immunized to tumor-specific antigens can adoptively transfer tumor-specific immunity to inbred normal animals.[1-5] When these treated animals are subsequently inoculated with the specific syngeneic tumor, inhibition of tumor development occurs. However, the adoptive transfer of tumor immunity by allogeneic leukocytes has had minimal success in man and experimental animals.[6-12] Immune elimination soon inactivates the transferred cells. Attempts at immunosuppression of the host to protect these cells may result in lethal graft-versus-host disease and have been uniformly unsuccessful.[13,14] Even the transfer of leukocytes compatible with the host at major histocompatibility loci has met with only limited success.[15]

Failure of allogeneic immune lymphocytes to prevent tumor development has led to an interest in the potential role of subcellular components, including the nucleic acids of these lymphocytes in transferring immune responses. The model for the immunotherapy of malignant disease with immune RNA was first suggested by Mannick and Egdahl[16] and became a reality with the work of Ramming and Pilch.[17] Mannick and Egdahl demonstrated that allograft immunity could be transferred to previously untreated rabbits by the administration of autologous lymphoid cells which had been preincubated with RNA extracted from the lymphoid organs of specifically immunized donor rabbits. Rabbits immunized by the adoptive transfer of RNA-incubated lymphocytes rejected specific skin grafts in an accelerated fashion, while indifferent third-strain skin allografts were rejected in the normal first-set fashion.[16,18] Pilch and Ramming adopted this model to the study of tumor immunity and demonstrated that immunity to murine tumor isografts could be transferred to previously untreated mice by the intraperitoneal administration of syngeneic spleen cells preincubated with RNA extracted from the lymph nodes and spleens of guinea pigs immunized with the specific mouse tumor being treated. Even though the guinea pigs were immunized to xenogeneic antigens as well as tumor antigens, the response was, indirectly at least, shown to be tumor specific.[19]

Deckers and Pilch then obtained direct evidence that immune RNA could mediate a specific immune response to tumor antigens alone by utilizing a system in which the tumor cell donor, the RNA donor, the spleen cell donor, and the recipient animal were all members of a single inbred strain.[20] This eliminated the possibility that the observed tumor isograft rejection might, in part, be due to an immune response directed against histocompatibility antigens and established conclusively the ability of spleen cells incubated with immune RNA to mediate destruction of tumor cells in vivo.

It had, therefore, been established that normal syngeneic spleen cells incubated with xenogeneic or syngeneic RNA extracted from the lymphoid organs of animals immunized with the tumor to be treated could be converted to a tumor-specific immunologic reactivity. These facts, in turn, suggested the possibility that lymphocytes sensitized to human tumor-specific antigens by incubation with RNA extracted from the lymphoid cells of animals immunized to these antigens could mediate a specific cytolytic effect on human tumor cells in vivo, thereby establishing a highly specific immunotherapy.[21]

However, no conclusive evidence of the regression of established murine or human tumors by treatment with immune RNA has been reported. We, and others, have repeatedly been able to adoptively transfer tumor-specific immunity using RNA from either syngeneic or

*Support by N.I.H. Grant CA15848-01, N.I.H. Contract A0143949, and N.I.H. GRS-NIH-RR05487-12 General Research Support Grant.

xenogeneic animals immunized with the tumor being studied.[19,20] In all such experiments, however, animals were treated with immune RNA-incubated lymphoid cells on the same day as the inoculation of the syngeneic tumor isograft. Although the animals were usually given a second inoculation of RNA-incubated lymphoid cells 24 and, occasionally, 48 hr later, this experimental model should not be construed as an immunotherapeutic attempt. Since the tumors were inoculated synchronously with the intraperitoneal injection of immune RNA, the inhibition of tumor development noted can only be interpreted as an example of transplantation resistance and not true immunotherapy. It remained to be demonstrated that immune RNA can be effective treatment of established viable growing tumor isografts which have been inoculated several days prior to the initiation of immunotherapy or which have become palpable tumors.

In addition, in all previous in vivo and in vitro studies, normal lymphoid cells were converted to an immunoreactive state by incubation with immune RNA preparations. Animals with growing tumors, however, ordinarily demonstrate a detectable immunologic reaction against the tumor-specific antigens of these tumors.[22] It is not clear that immune RNA can be effective in augmenting the tumor-specific immunologic reactivity of lymphocytes from animals with growing tumors. Recent experiments in this laboratory have been directed toward answering these two questions concerning RNA-mediated transfer tumor immunity.

For these experiments, inbred male and female C3H/HeN mice were used. The tumors employed were sarcomas, designated BP-8 and BP-9, and were induced in adult female C3H/HeN mice by the subcutaneous injection of 0.1 ml of a 1% solution of 3,4-benz(a)pyrene in sesame oil. Both BP-8 and BP-9 sarcomas were maintained by serial transplantation in adult female C3H/HeN mice and were used in their third through tenth transplant generations.

Single-cell suspensions were prepared from these tumors by a modification of the method of Hammond et al.[23] Tumor tissue was minced and placed in trypsinizing flasks. Dulbecco's solution containing 0.25% pronase was then added (10 ml/g of tumor tissue). The mixture was agitated by a magnetic stirrer for 10 to 15 min at room temperature and filtered through sterile cheesecloth. Then, 0.04% DNase (0.3 ml/g of tumor tissue) was added to the suspension. The cell suspension was centrifuged at 250 $\times$ g for 10 min at 4°C, resuspended in Hanks' balanced salt solution (BSS) (containing penicillin and streptomycin), and washed three times in BSS. The final cell button was resuspended in BSS and viable cells quantitated in a hemocytometer by determining the number of cells which excluded 0.4% trypan blue. Approximately 90 to 95% of the cells in these tumor-cell suspensions were viable by this criterion. The cumulative incidence of tumor development following the subcutaneous inoculation of varying numbers of viable BP-8 and BP-9 tumor cells in untreated female C3H/HeN mice was then determined.

The presence of tumor-specific transplantation antigens in the BP-8 and BP-9 sarcomas was established after the manner of Prehn and Main.[24] Two groups of untreated mice were inoculated in a hind limb with 10^5 BP-8 or BP-9 cells, respectively. Tumors developed in all animals; when these tumors were approximately 1 to 1.5 cm in transverse diameter, the hind limbs of these animals were amputated. Seven days later these amputated animals were challenged with either BP-8 or BP-9 tumor cells. This second challenge inoculum represented a dose of tumor cells calculated to develop progressively growing tumors in 80 to 90% of injected animals and was approximately 10^3 to 10^4 tumor cells in 0.1 ml of tissue culture media. In this study no animal which had had a growing BP-8 tumor amputated and was then rechallenged with BP-8 tumor cells developed a second viable tumor isograft. Similar results were obtained for animals immunized to the BP-9 tumor by amputation of growing BP-9 isografts. In addition, it was demonstrated in these studies that an inoculum of 10^3 tumor cells of either the BP-9 or the BP-8 tumor resulted in the development of tumor isografts in approximately 80 to 85% of normal animals. Similarly, in other studies, we have demonstrated that these two chemically induced fibrosarcomas contain distinct tumor-specific transplantation antigens without detectable cross reactivity.* The antigenicity and immunogenicity, therefore, of these two tumors

*Unpublished data.

within their murine strain of origin were conclusively demonstrated.

Hartley guinea pigs were immunized to either one of these two tumors in the following manner. A brei of tumor cells was prepared by forcing minced murine tumor tissue suspended in Hanks' BSS in a ratio of 40% w/v through a #40-mesh stainless-steel screen. The guinea pigs were immunized by injecting into each footpad a mixture of 0.25 ml of mouse tumor cell brei and an equal volume of complete Freund's adjuvant for a total injection volume of 0.5 ml per footpad. An intraperitoneal injection of 1 ml of tumor cell brei without adjuvant was also given. Control guinea pigs received similar injections of mouse normal tissues (a mixture of lung, liver, kidney, and spleen cells) in complete Freund's adjuvant. Other control groups of guinea pigs received only injections of Freund's adjuvant. After 10 to 14 days, the spleens and axillary, popliteal, and inguinal lymph nodes from all groups of guinea pigs were excised, frozen in dry ice, and the RNA extracted.

From some in vitro studies, C3H/HeN mice were immunized by progressive tumor growth. We have previously demonstrated that mice with growing tumors develop a tumor-specific immunologic reaction against these tumors and that tumor-specific transplantation immunity can be transferred to previously normal, untreated mice by intraperitoneal inoculation of lymphoid cells from tumor-bearing mice.[22,25] The adoptive transfer of tumor immunity in this fashion is fully developed when lymphoid cells are removed from tumor-bearing mice 14 days after inoculation of the viable tumor isografts; it is lost at some indefinite period when the tumors of the lymphoid cell donor become excessive in size. This phenomenon is referred to as concomitant tumor immunity; it demonstrates that animals with growing tumors are immunized to the tumor-specific transplantation antigens of these tumors and that this immunity resides in the lymphoid cells of these animals.

For RNA extraction, homogenization of the guinea pig lymphoid tissues was performed in a Sorvall® omnimixer* containing phenol which had been washed three times with a 0.01 *M* sodium acetate buffer containing 0.1% 8-hydroxyquinolone** and an equal volume of 0.01 *M* acetate buffer pH 5.0 containing 0.1% sodium dodecyl sulfate and 2 mg/ml of polyvinyl sulfuric acid potassium salt.*** Ten drops of bentonite solution† (11.4 mg/ml) was also added to this mixture. Lymphoid tissues were homogenized at 4° C for 3 min, transferred to a 40-ml centrifuge tube, and heated to 50° C in a water bath with constant stirring. The mixture was then immediately cooled to 4° C in an ice bucket containing alcohol and dry ice and centrifuged at 28,000 × g at 4° C for 10 min. The aqueous phase containing the RNA was carefully removed with a pipette. An equal amount of buffer-washed phenol was added to the aqueous phase and the extraction procedure repeated three times. After the last extraction, the aqueous phase was adjusted to 0.3 *M* with 2.4 *M* acetate buffer and the RNA precipitated with absolute ethanol. The precipitate was collected by centrifugation at 1,400 × g at 4° C for 5 min and washed twice in 30 ml of 75% ethyl alcohol. The pellet was then dissolved in Hanks' solution without phenol red. The RNA preparations used had an OD:260-to-OD:280 ratio of 2.0 or greater in a Beckman DU® spectrophotometer.

In addition, 0.7-ml aliquots of the various RNA preparations were carefully layered on a 5 to 20% linear continuous sucrose gradient in cellulose nitrate tubes. The tubes were centrifuged in a Spinco SW-39L rotor at 38,000 rpm at 4° C for 4 hr in a Beckman L3-50 ultracentrifuge. Ten-drop samples were collected by a density-gradient fractionator* and examined optically at 260nm on a Beckman DU spectrophotometer. Sucrose density-gradient fractionation of the lymphoid RNA generally yielded three peaks, as determined by optical density at 260 nm (Figure 1). These peaks represented approximate sedimentation coefficients of 28, 18, and 4S, respectively. If the 28S peak was larger than the 18S peak, which in turn was larger than the 4S peak, the RNA was considered nondegraded and was used for subsequent incubation with lymphocytes.

Spleens were excised from mice immunized by

*Fisher Scientific Company, Fairlawn, New Jersey.
**Matheson, Coleman and Bell, East Rutherford, New Jersey.
***Eastman Organic Chemicals, Rochester, New York.
†Buchler Instruments, Fort Lee, New Jersey.

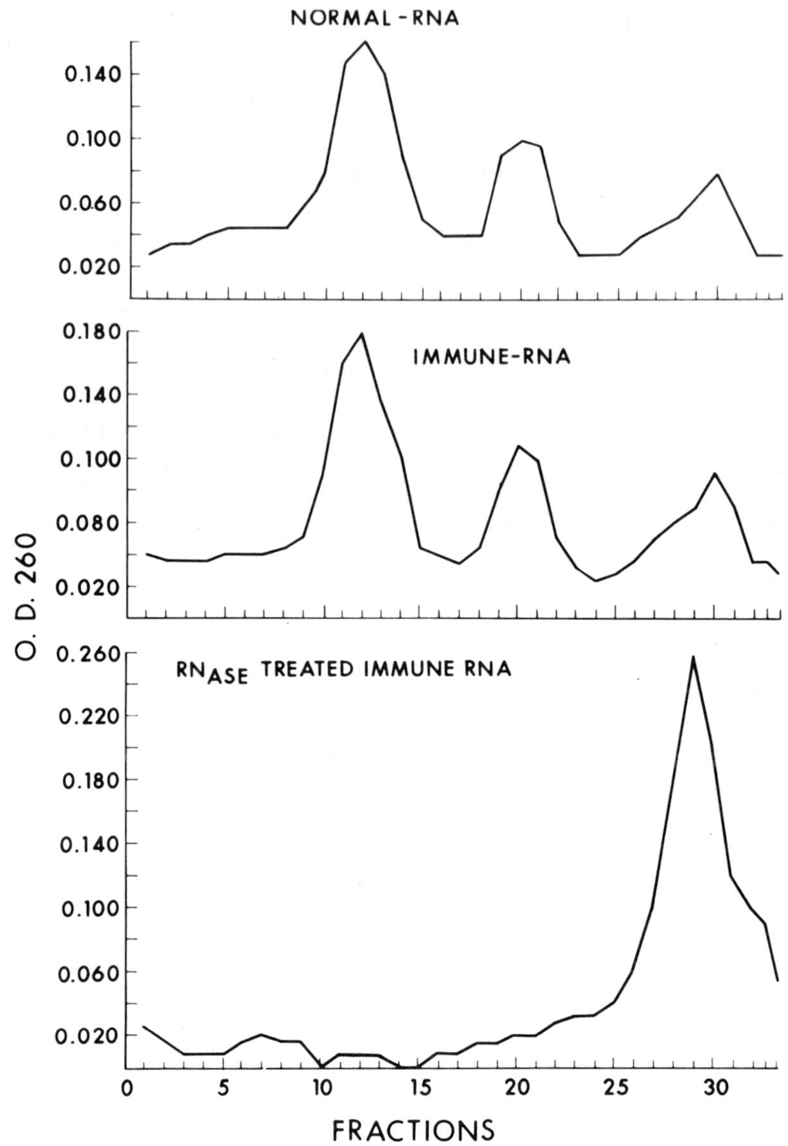

FIGURE 1. Sucrose density gradient profile of a representative immune RNA. Three peaks with approximate sedimentation coefficients of 28, 18, and 4S, respectively, from left to right are shown.

progressive growth of a tumor isograft. In addition, spleens were also removed from normal C3H/HeN mice. These spleens were individually minced with scissors in cold BSS and passed through #40-mesh stainless-steel screens. The suspension was then filtered through #80-mesh stainless-steel screens and centrifuged at 200 × g for 10 min. The cells were washed once with cold BSS. For incubation with RNA, the cell button was resuspended in the RNA solution at a concentration of 10^7 to 10^8 cells/ml. RNA concentrations were always 500 to 1,000 μg/ml. The mixtures of RNA and spleen cells were then incubated at 37° C for 20 min in a gyratory shaking water bath, washed once with cold BSS, counted, and suspended in RPMI 1640 with 20% fetal calf serum at a concentration of 4×10^7 viable lymphoid cells per milliliter. Viability was determined by trypan blue exclusion. In addition, for the in vitro studies, purification of the lymph-

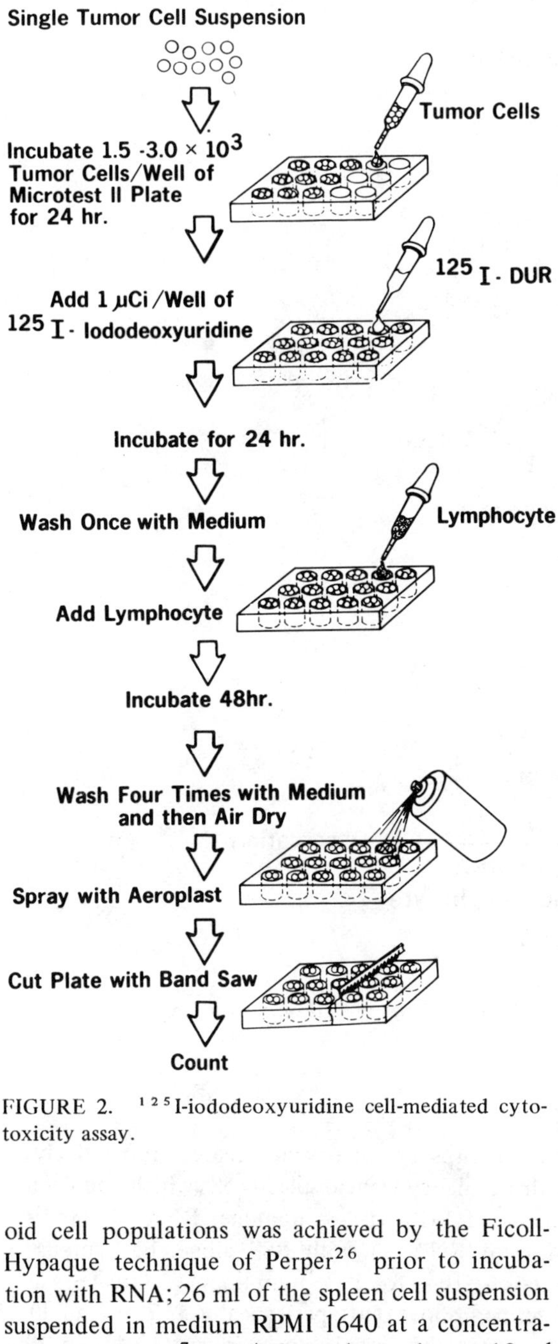

FIGURE 2. ^{125}I-iododeoxyuridine cell-mediated cytotoxicity assay.

oid cell populations was achieved by the Ficoll-Hypaque technique of Perper[26] prior to incubation with RNA; 26 ml of the spleen cell suspension suspended in medium RPMI 1640 at a concentration of 3 X 10^7 cells/ml was layered over 10 ml of the Ficoll-Hypaque solution: a mixture of one part of 50% (w/v) sodium diatrizoate (Hypaque) with four parts of 8% (w/v) Ficoll in an 11.5 X 2.5 cm round-bottomed glass tube. Tubes were centrifuged at 400 X g in a swinging bucket rotor at 4° C for 25 min. Following centrifugation, the formed elements had layered so that the erythrocyte and granulocyte layers were at the bottom of the tube, the gradient fluids next, and the mononuclear lymphocytes and a few platelets delivered to the interface between the medium and the gradient fluids. This layer was 2 cm above the nearest cell layer (granulocytes) and could be removed without contamination. An examination of the recovered mononuclear cell layer revealed a population of greater than 95% lymphocytes with a viability of greater than 98% as judged by trypan blue exclusion. These purified lymphocytes were used in the in vitro cytotoxicity assay to be described. Prior to incubation with immune RNA, they were washed at least three times with tissue culture medium.

A modification of the cell-mediated cytotoxicity assay using ^{125}I-iododeoxyuridine (^{125}I-DU)-labeled target cells, as originally described by Cohen,[27] was used for these studies. Single-cell BP-8 or BP-9 tumor cell suspensions were prepared as described above. The assay was then performed, as described below and depicted in Figure 2. Approximately 2,000 BP-8 and BP-9 cells in 0.2 ml, of RPMI 1640 containing 1% glutamine, 100 µ/ml of penicillin, 100 µg/ml of streptomycin, and $10^{-6}M$ fluorodeoxyuridine (FUDR)/ml were then added to each well of a Microtest II® culture plate.* One microcurie of ^{125}I-DU was also added to each well, and the cultures were incubated at 37°C in a 10% CO_2 water-saturated atmosphere for 24 hr. Excess ^{125}I-DU was then aspirated, and each well was washed once with RPMI 1640. Lymphocytes, in 0.2 ml of RPMI 1640, were then added. In some experiments, these lymphocytes were normal, previously untreated lymphocytes removed from mice never immunized to either the BP-8 or BP-9 tumor or were normal lymphocytes which had been incubated with various RNA preparations. In still other experiments, these were immune lymphocytes removed from animals immunized by progressive growth of a tumor isograft or immune lymphocytes which had been preincubated with RNA extracted from the lymphoid organs of animals immunized to either the BP-8 or BP-9 tumor as described previously.

After 48 hr of incubation, lymphocytes and killed target cells were removed by washing, and the remaining radioactive adherent target cells were sprayed with aeroplast. Each well was then

*Falcon Plastics, Oxnard, California.

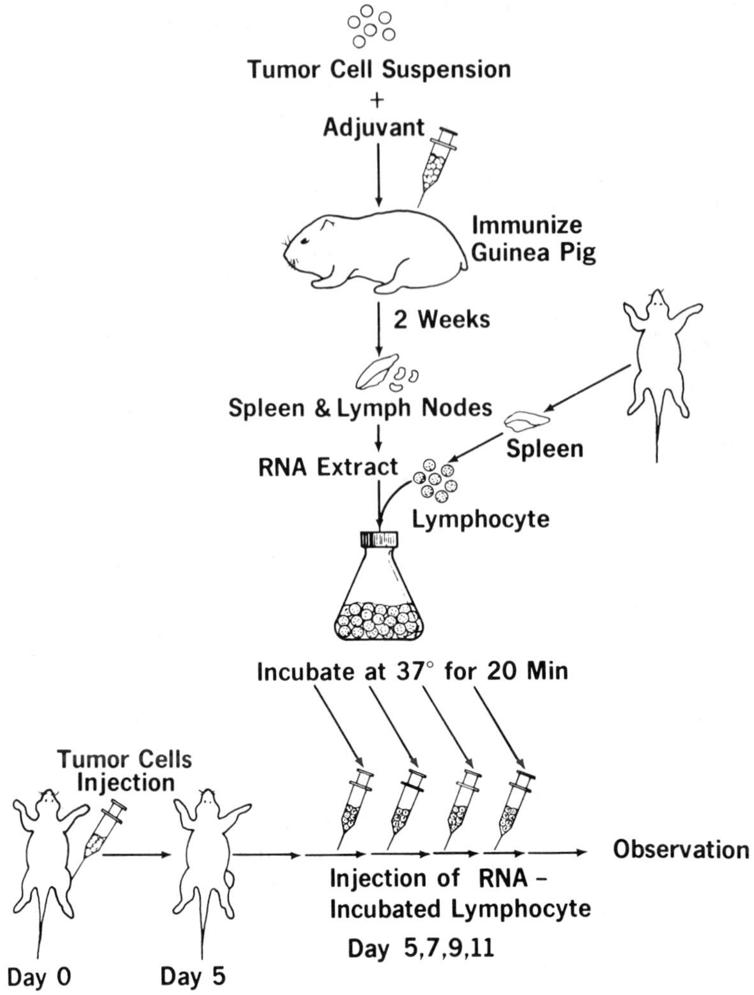

FIGURE 3. Model for the immunotherapy of BP-8 sarcomas in C3H mice with immune RNA-incubated lymphocytes.

cut with a band saw, and radioactivity was measured in a gamma counter. Five to eight replicate wells were tested in each group, and a mean counts per minute (cpm) together with the standard deviation of each cpm was calculated for each group.

The in vivo immunotherapy model is described below and depicted in Figure 3. Hartley guinea pigs were immunized with either the BP-8 or BP-9 tumor, with Freund's adjuvant alone, or with normal mouse tissues and used as donors of immune RNA. In some experiments, BP-8 or BP-9 immune RNA was degraded with RNAse prior to incubation with normal mouse spleen cells. Four groups of C3H/HeN mice were inoculated on day 0 with 10^3 viable BP-8 tumor cells. Five days later, these groups of mice were treated, respectively, with normal syngeneic spleen cells which had been incubated with BP-8 immune RNA, Freund's adjuvant RNA, or with BSS alone. Each mouse received 10^7 to 10^8 lymphocytes (treated as described above) intraperitoneally 5, 7, 9, and 11 days after the initial tumor isograft. Mice in all groups were then observed for tumor development. Two times each week tumor size for all the mice in each group was calculated, and a mean tumor volume for all the animals so studied was then calculated by the formula of Attia and Weiss[28] where volume = $0.4\ ab^2$. In a second similar experiment, four groups of syngeneic C3H/HeN mice were inoculated on day 1 with 10^3 BP-8 tumor cells. Seven days later these animals

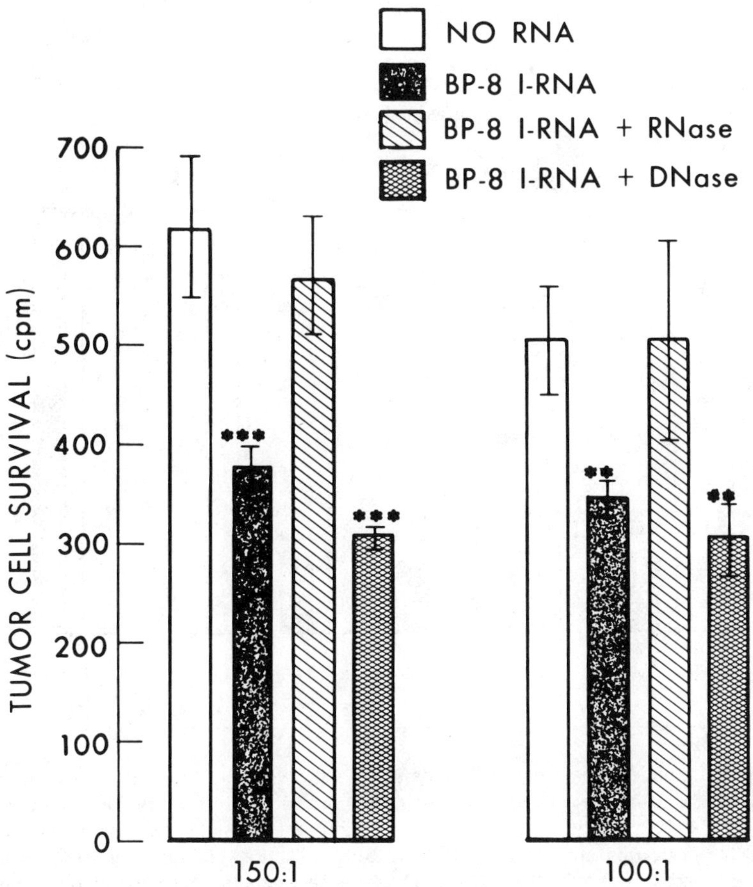

FIGURE 4. Immune cytolysis of BP-8 tumor cells mediated by the lymphoid cells of syngeneic normal mice preincubated with xenogeneic BP-8 immune RNA, BP-8 immune RNA-RNase, BP-8 immune RNA-DNase, or with phosphate-buffered saline alone (no RNA). (***) $P<0.001$ by Student's t test when cytolysis observed in experimental groups is compared to cytolysis observed with normal lymphoid cells alone. (**) $P<0.01$ when cytolysis observed in experimental groups is compared to cytolysis observed with normal lymphoid cells alone. Similar survival observed for untreated tumor cells and normal lymphocytes alone.

were treated by the intraperitoneal inoculation of 10^7 to 10^8 viable syngeneic lymphocytes which had been preincubated with BP-8 immune RNA, BP-8 immune RNA previously degraded with RNase (15 μg/ml of RNA), or BP-9 immune RNA. Again, all animals were followed for tumor development. The size of each tumor was calculated as described previously.

Our initial in vitro experiments were designed to test whether normal, nonimmune syngeneic C3H/HeN mouse spleen cells when preincubated in vitro with xenogeneic immune RNA extracted from the lymphoid organs of guinea pigs immunized with BP-8 or BP-9 tumor cells could be converted into cells specifically cytotoxic to BP-8 or BP-9 tumor cells in vitro. The results of these experiments are given in Figure 4. At lymphoid cell-to-target cell ratios of 100:1 and 150:1, normal C3H/HeN spleen cells preincubated with BP-8 immune RNA caused cytolysis of BP-8 target cells. Statistically significant cytotoxicity was observed at all lymphoid cell-to-target cell ratios. In addition, significant cytolysis of BP-8 tumor cells by normal lymphocytes was never observed. Furthermore, the pretreatment of immune RNA with RNase resulted in degradation of the RNA and abrogation of the immune cytolysis observed, indicating the necessity for intact immune RNA in

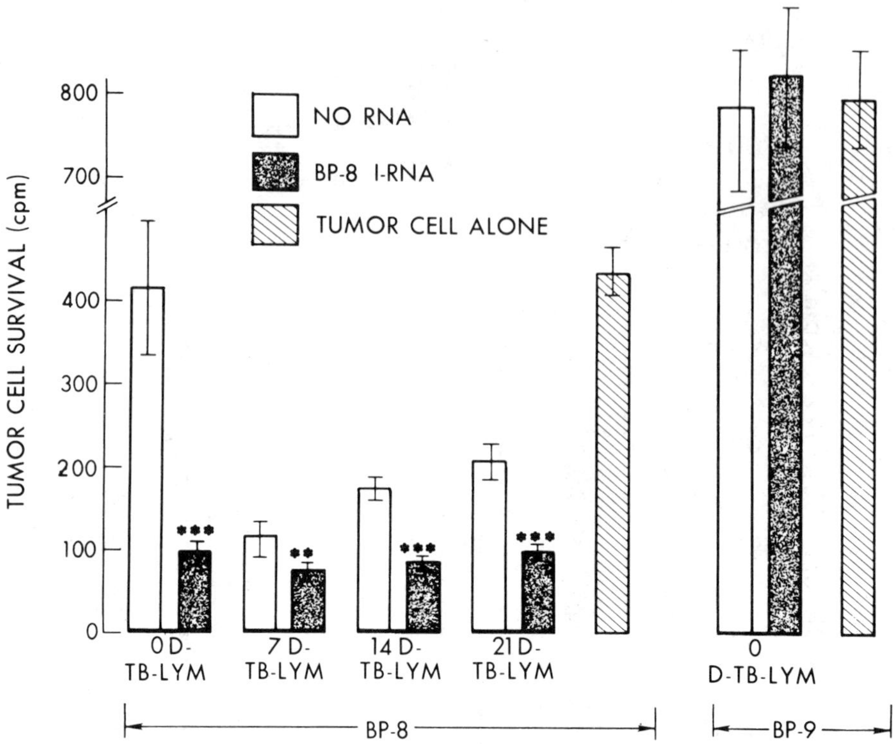

FIGURE 5. Immune cytolysis of BP-8 tumor cells mediated by the lymphoid cells of BP-8 tumor-bearing mice and the augmentation of this observed cytolysis by preincubation of these BP-8 immune lymphoid cells with xenogeneic BP-8 immune RNA. The activity of lymphocytes from BP-8 tumor-bearing mice against BP-8 and BP-9 cells at several intervals after inoculation of the BP-8 isograft is shown (no RNA). (**) $P<0.01$ when the cytolysis of lymphoid cells observed after incubation with BP-8 immune RNA is compared to that observed prior to incubation. (***) $P<0.001$.

these experiments. However, treatment with DNase failed to change the immune cytolysis observed. This indicates most likely that DNA is not an integral component of the active material involved in converting nonimmune cells to states of active immunity.

In all previous experiments, lymphoid cells were removed from normal mice and then incubated with appropriate RNA solutions before they were tested for their cytotoxic effects on tumor cells. In a separate experiment, mice were inoculated with BP-8 tumor on day 0, then 7, 14, and 21 days later these mice were sacrificed and their spleen cells removed. These lymphoid cells were either directly tested in vitro against BP-8 tumor cells or were first preincubated with xenogeneic BP-8 immune RNA extracted from the lymphoid organs of guinea pigs immunized with BP-8 tumor. The results (Figure 5) show that on day 0 lymphoid cells from BP-8 tumor-bearing animals demonstrated very little cytotoxicity toward BP-8 cells. However, incubation of aliquots of the cell suspension with BP-8 immune RNA resulted in the appearance of a statistically significant cytotoxic effect to BP-8 cells compared to cells from animals not exposed to RNA. At 7, 14, and 21 days after the initial tumor inoculation, the lymphoid cells of tumor-bearing mice were specifically cytotoxic to BP-8 tumor cells in vitro, independent of any preincubation with immune RNA. However, incubation of these tumor-specific effector lymphoid cells with BP-8 immune RNA extracted from the lymphoid organs of guinea pigs immunized with the BP-8 tumor resulted in a significant augmentation of this immunoreactivity relative to control groups. These data suggest that the cytolytic capacity of the immune response is not quantitatively maximal in the tumor-bearing animal and that this capacity can be augmented by preincubation with immune RNA. Whether

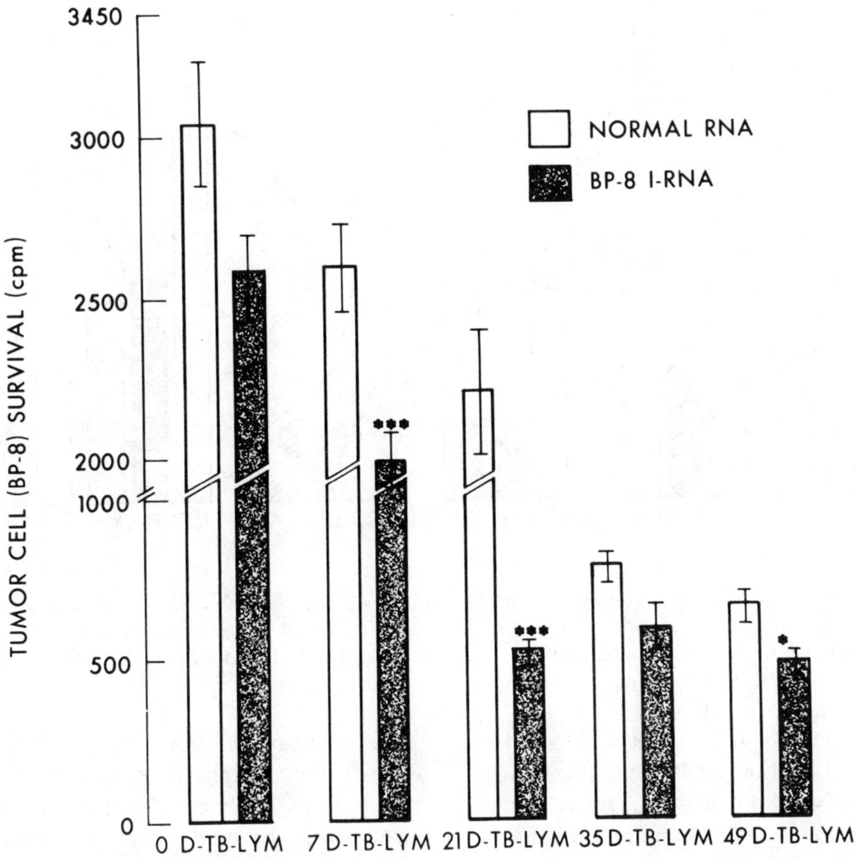

FIGURE 6. Immune cytolysis of BP-8 tumor cells mediated by the lymphoid cells of BP-8 tumor-bearing mice preincubated at several intervals after initial tumor inoculation with either BP-8 immune RNA or normal tissue RNA. At each interval, nonincubated lymphoid cells from BP-8 tumor-bearing mice produced cytolysis similar to that observed for lymphoid cells preincubated with normal tissue RNA. (*) $P<0.05$. (***) $P<0.001$.

"uncommitted" cells are stimulated to respond or whether "committed" cells respond more "vigorously" after exposure to RNA extracts was not determined. In addition, in this experiment it was determined that lymphoid cells from BP-8 tumor-bearing mice had no detectable cytotoxic capacity against monolayers of BP-9 tumor cells. Similarly, incubation of lymphoid cells from BP-8 tumor-bearing mice with BP-8 immune RNA did not lead to cytolysis of BP-9 tumor cells in vitro, supporting the contention that the BP-8 and BP-9 tumors are immunogenic in their murine strain of origin and that their unique antigenic characteristics are non-cross reacting. As these observed immune responses could have been directed against mouse transplantation antigens, experiments were designed to further investigate the specificity of these reactions toward different tumors. Spleen cells from mice inoculated 1, 7, 21, 35, or 49 days earlier with viable BP-8 isografts were preincubated with RNA extracted from the lymphoid organs of guinea pigs immunized with normal C3H/HeN spleen cells or with BP-8 tumor tissue. Data from these experiments are recorded in Figure 6. The results indicate that BP-8 immune RNA specifically augmented the immunoreactivity of lymphoid cells from mice bearing BP-8 tumors. RNA extracts of normal tissue were not effective in this regard. At each interval tested, the cytolysis produced by lymphoid cells treated with normal tissue RNA was similar to that produced by the lymphoid cells alone.

In the experiment shown in Figure 7, mice were inoculated with BP-8 tumor cells on day 0 and tumors were allowed to develop through 5 weeks of normal tumor growth. Tumor-bearing mice

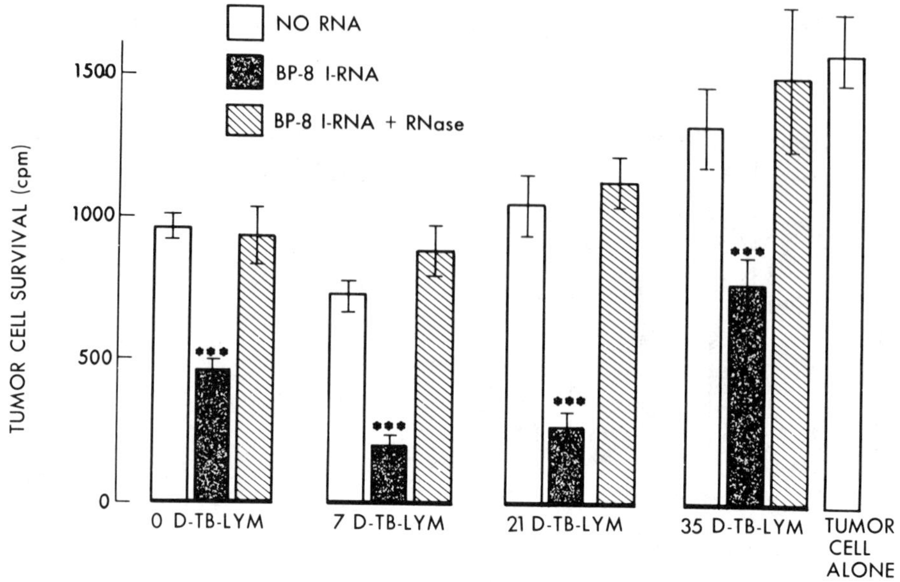

FIGURE 7. Immune cytolysis of BP-8 tumor cells by the lymphoid cells of BP-8 tumor-bearing mice preincubated at several intervals after initial tumor inoculation with BP-8 immune RNA, BP-8 immune RNA-RNase, or with phosphate-buffered saline (no RNA). (***) $P<0.001$.

were sacrificed on days 0, 7, 21, and 35 of this experiment, and lymphoid cells from these mice were tested at a lymphoid cell-to-target cell ratio of 250:1 against BP-8 tumor cells in vitro. Lymphoid cells from these mice were specifically immunoreactive against BP-8 tumor cells 7 and 21 days after inoculation of the BP-8 tumor, but by 35 days after inoculation this immunoreactivity had been lost. However, 0, 7, 21, and 35 days after inoculation with the BP-8 tumor cells, lymphoid cells from these mice when mixed with BP-8 immune RNA exhibited cytotoxic effects against BP-8 tumor cells in vitro. In addition, at each interval studied, treatment of the BP-8 RNA with RNase prior to incubation with BP-8 immune lymphocytes resulted in complete abrogation of the observed response and a return to background reactivity. These data again indicate that concomitant tumor immunity can be augmented by incubation of tumor-specific immune lymphocytes with specific immune RNA and that intact immune RNA is necessary for the augmentation to occur. In addition, it confirms the observation that in the presence of a large tumor mass the immunoreactivity of the host against the growing neoplasm may be lost. Significantly, the data suggest that lost immunoreactivity may be restored by incubation of these lymphocytes with specifically immune RNA before testing.

As mentioned above, to study the effect of RNA-treated lymphoid cells on tumor growth in vivo, groups of Hartley guinea pigs were immunized with Freund's adjuvant alone, Freund's adjuvant mixed with BP-8 tumor cells, Freund's adjuvant mixed with normal mouse tissue, and Freund's adjuvant mixed with antigenically distinct BP-9 tumor cells. Ten to fourteen days after these immunizations, all groups of animals were sacrificed. RNA from lymphoid tissues was extracted as described previously. As before, all experimental mice were inoculated with 10^3 BP-8 tumor cells on day 0; 5, 7, 9, and 11 days later these animals received intraperitoneal inoculations of previously untreated syngeneic C3H/HeN lymphocytes incubated previously with BP-8 immune RNA, BP-9 immune RNA, RNA from normal tissues, Freund's adjuvant RNA, or BP-8 immune RNA degraded previously during incubation with RNase. In other words, immunotherapy was begun 5 days after the injection of the tumor isograft and at a time when the tumor isograft was, presumably, growing actively. Other control groups included tumor-bearing animals receiving tumor cells alone and animals which received only normal spleen cells as a method of immunotherapy. The results (Table 1) indicated some decrease in the incidence of tumor development early in the experiment, and this delay in tumor

TABLE 1

Immunotherapy of Established Murine Tumors by Spleen Cells Preincubated with RNA

	Tumor incidence[a] days after tumor isografts					
Experimental groups	20	23	26	28	30	40
Tumor cells alone (A)	6/20	8/20	12/20	14/20	16/20	16/20 (471)[b]
Spleen cells (no RNA) (B)	8/17	9/17	14/17	14/17	15/17	15/17 (279)
BP-8 RNA (C)	0/16	3/16	6/16	9/16	13/16	13/16 (429)
Freund's RNA (D)	4/13	6/13	9/13	9/13	10/13	10/13 (432)

[a]Cumulative incidence of tumor development expressed as number of mice with tumors/ number of mice in the group.

[b]Number in parentheses indicates mean tumor volume (cm^2), volume = $0.4\ ab^2$ where a is the major axis and b the minor axis of the tumor at right angles.

appearance was statistically significant. Eventually, however, the incidence of tumor development equalized in all groups, and, although initially there was a striking decrease in the size of the palpable tumors in the animals treated with cells exposed to BP-8 immune RNA, eventually the difference in tumor size between these groups was no longer demonstrable. In a second experiment, the specificity of the effect, as well as the necessity for intact immune RNA, was found. Again, animals were treated on days 5, 7, 9, and 11 and observed for tumor development. As can be seen in Table 2, the results were similar to those observed in the first immunotherapy experiment. Mice receiving treatment with BP-8 immune RNA-incubated lymphoid cells had an early statistically significant decrease in tumor development. At this early point in the experiment, these observations were tumor specific. Eventually, however, all animals again developed tumors, and the growth rate of the tumors equalized in all of these experiments.

DISCUSSION

Tumor-specific transplantation immunity can be induced experimentally in syngeneic hosts against a wide variety of chemical- and virus-specific neoplasma[29-31] This immunity is usually detected by the rejection of tumor isografts in animals immunized by excision of growing transplants of the tumor under investigation. In addition, the lymphoid cells of animals so immunized can adoptively transfer tumor-specific immunity against subsequent tumor isografts to normal animals. However, these facts provide no direct information about the immunological response of a host to its tumor prior to experimental interference with the tumor, since immunity was produced by procedures which prevented tumor growth.

However, much recent evidence suggests that tumor-specific immunological response can be observed in most animal tumor-host systems, despite the relentless growth of the immunizing tumor isograft. Indeed, we have repeatedly found that a primary tumor isograft grows progressively, despite an immune response that prevents cellular aliquots of the same tumor from growing elsewhere in the tumor-bearing host.[32] We have also determined, in a syngeneic murine tumor-host system, that the adoptive transfer of spleen cells from mice with growing tumors to previously untreated syngeneic mice prevents the growth of an isograft of the same tumor in the spleen cell recipients.[22]

In addition, other investigators have found, using in vitro tests of tumor immunity, that progressive growth of a primary tumor does not necessarily indicate the absence of an antitumor immune response. Rosenau and Morton reported that lymphoid cells from mice bearing transplants of a MCA-induced sarcoma suppressed the growth of explants of the immunizing tumor in vitro.[33] Barski and Youn also noted that peritoneal exudate cells removed from Balb/c mice bearing tumors induced by an attenuated Rauscher virus inhibited colony formation of these tumor cells in vitro.[34]

Recently, the existence of tumor-specific cell

TABLE 2

Immunotherapy of Established Murine Tumor by Spleen Cells Preincubated with RNA

	Tumor incidence[a] days after tumor incidence						
Experimental groups	14	16	19	23	26	29	40
Tumor cells alone	6/23	8/23	14/23	18/23	19/23	20/23	20/23 (377)[b]
BP-8 RNA	0/22	5/22	7/22	13/22	13/22	14/22	14/22 (212)
BP-8 RNA-RNase	4/10	4/10	6/10	7/10	7/10	8/10	8/10 (224)
BP-9 RNA	6/21	10/21	11/21	14/21	15/21	16/21	16/21 (339)

[a]Cumulative incidence of tumor development expressed as number of mice with tumors/ number of mice in the group.
[b]Number in parentheses indicates mean tumor volume (cm^2), volume = $0.4\ ab^2$ where a is the major axis and b the minor axis of the tumor at right angles.

surface antigens in a variety of human tumors has been confirmed,[35-41] and evidence exists to suggest that patients with cancer mount a significant immune response to tumor-specific antigens despite the persistent growth of the primary and metastatic tumor.[35,40-42]

These immune responses directed against tumor-associated antigens and detected in a host bearing a progressively growing tumor are referred to as concomitant tumor immunity. The effective immunotherapy of cancer may depend on understanding the paradox of progressive growth of antigenic tumor in an immunologically competent host, despite the concomitant existence of immune responses against the tumor-specific antigens of that tumor.

The existence of concomitant tumor immunity has interesting implications for those who advocate the adoptive transfer of immune sera, cells, or subcellular fractions as a method of passive immunotherapy. Indeed, in most animal tumor-host systems, it has been accepted that cell-mediated immune responses are primarily responsible for tumor cell destruction and that humoral antibodies play a secondary role. Humoral antibodies may, in fact, enhance tumor growth because of their ability to inhibit cellular immunity by directly "blocking" the sensitized lymphocytes' ability to interact with and kill tumor cells.[37,42] However, humoral factors are not sufficient explanation for the protection of tumor cells from destruction by a tumor-specific cell-mediated immunological response which is operative during tumor growth. The existence of concomitant tumor immunity in vivo, defined as the ability of the tumor-bearing host to reject a second isograft of the same tumor, is incompatible with the postulated presence of significant humoral tumor-specific blocking factors, which should, by definition, protect the second isograft as well as the first one. Indeed, though the existence of humoral factors capable of blocking tumor destruction in vitro has been demonstrated in several experimental and human tumor-host systems,[37,43] we would suggest that enthusiasm for this hypothesis must be tempered by the realization that it cannot explain the phenomenon of concomitant tumor immunity in all experimental systems.

Much experimental evidence implies that the growth of the primary tumor may represent a relative inadequacy of the cell-mediated immune response. The proliferative capacity of the viable tumor cells may exceed the tumor-specific cytolytic ability of specifically sensitized lymphocytes. This viewpoint accepts the belief that the problem of continued tumor growth despite the presence of a strong antitumor immune response is one primarily of either logistics (inadequate numbers of sensitized cells) and/or a quantitative inefficiency of adequate numbers of sensitized cells. Passive immunotherapy attempts to increase the numbers of sensitized cells and/or augment their killing efficiency.

In these experiments, we found that normal nonimmune murine lymphoid cells can be converted to effector cells specifically cytotoxic to syngeneic murine tumor cells by incubation in vitro with xenogeneic immune RNA extracted from the lymphoid tissues of specifically immunized guinea pigs. This response has been shown to be specific for the tumor used to immunize the RNA donor. Spleen cells preincubated with RNA extracted from the lymphoid organs of guinea pigs immunized with the BP-8 tumor cells were cytotoxic only to BP-8 target cells and not to target cells of the syngeneic, but antigenically distinguishable, BP-9 tumor. Nonspecific factors associated with guinea pig lymphoid RNA or immune responses directed against mouse transplantation antigens could not account for the cytotoxic activity acquired by normal spleen cells after incubation with BP-8 immune RNA since RNA preparations extracted from the lymphoid organs of guinea pigs immunized with normal C3H/HeN tissues failed to mount an antitumor immune response. This evidence strongly suggests that immune responses mediated by this type of xenogeneic immune RNA preparation are directed specifically against tumor-associated antigens. In addition, we have observed that active immune RNA preparations treated with ribonuclease are inactivated, whereas treatment with deoxyribonuclease is without effect, indicating that RNA is involved in our immunoreactive nucleic acid preparations. Utilizing an inbred syngeneic murine tumor-host system, we have also found that ^{125}I-DU cytotoxicity assay is useful in determining, quantitatively, the immunoreactivity of lymphocytes from animals with growing tumors against monolayers of that same tumor. Using this system, we have demonstrated that preincubation of lymphoid cells from animals with growing tumors with xenogeneic immune RNA, extracted from the lymphoid organs of animals immunized with the murine tumor being studied, is capable of significantly augmenting the immunoreactivity of these lymphoid cells against specific tumor cells in vitro. Concomitant tumor immunity can be augmented by incubation of already committed specifically sensitized lymphoid cells with tumor-specific immune RNA. This finding, as discussed previously, is clearly important if RNA is to be employed clinically as a means of immunotherapy. Obviously, animals and patients with growing tumors may have intact immune responses against these particular tumors, and one problem may well be a quantitative inadequacy of their lymphocyte-mediated tumor-specific immune response. These data confirm the fact that immune RNA can augment the efferent arc of a tumor-specific immune response. They make it logical to proceed with this modality as a method of passive immunotherapy of established rodent and human tumors.

In addition, we have presented our initial experiments in the immunotherapy of rodent tumors with immune RNA. In these early experiments, a statistically significant delay in tumor development was observed, and this event was shown to be specific for the immunizing tumor and to be independent of any nonspecific effect of murine lymphoid cells themselves. In addition, it was found that intact immune RNA was necessary for this effect. However, in both of these early immunotherapy experiments, despite treatment beginning only 5 days after inoculation of the viable tumor isograft and proceeding every other day through the next week, no statistically significant decrease in final tumor incidence or in the growth of these tumors was ever seen. Nevertheless, these preliminary experiments suggest that immune RNA may be a successful method for the immunotherapy of established tumors, and experiments are currently underway to determine the effect of multiple treatment courses with immune RNA-incubated lymphoid cells and to test the possibility that direct injections of immune RNA with or without an RNase inhibitor may be effective in eradicating established rodent tumors.

Evidence to support use of this more direct and simpler approach was first presented by Alexander who extracted xenogeneic immune RNA from the lymph cells of sheep immunized with "chunks" of primary benz(a)pyrene-induced rat sarcomas and injected these RNA preparations into the footpads of tumor-bearing rats. Administration of these immune RNA preparations resulted in tumor growth retardation and occasional complete regression of tumors in rats bearing the same sarcoma used to immunize the RNA-donor sheep.[44] However,, in most immunoprophylactic and immunotherapeutic models, direct injections of RNA have not been effective, presumably because of inactivation of immune RNA by interstitial and plasma RNases. Deckers and Pilch circumvented this problem by administering their RNA in solutions containing a potent RNase inhibitor (sodium dextran sulfate) and observed, in a syngeneic murine as well as a syngeneic rodent tumor-host system, that animals treated with tumor-specific xenogeneic immune RNA mixed with dextran sulfate demonstrated significant inhibition of tumor development compared to all control groups.[45,46] A similar immunoprophylactic effect of direct injections of immune RNA was reported by Kennedy et al.[47] They prepared allogeneic immune RNA to a murine tumor and injected this RNA into the footpads of untreated mice in phosphate-buffered saline (pH 6.9) containing 0.1 mg/ml of polyvinyl sulfate, a potent RNase inhibitor. Mice so treated were rendered resistant to tumor isografts. Recently, Pilch has found that the successful immunotherapy of a transplantable, spontaneously metastasizing, chemically induced mammary adenocarcinoma in female Fischer 344/4 rats was possible using direct injections of xenogeneic immune RNA mixed with the RNase inhibitor, sodium dextran sulfate. He has initiated Phase I trials of the immunotherapy of cancer in man with direct injections of immune RNA.[48]

Although not of apparent immediate clinical importance, the exact method of action of immune RNA in the transfer of antitumor immunity has aroused considerable scientific interest. Its relationship, if any, to the transfer factor described originally by Lawrence and coworkers[49,50] has not been established.

Two types of immune RNA have been described. One is a highly immunogenic complex of a unique species of RNA with antigen or antigenic determinants and may behave as a "super antigen." The second type is free of detectable antigenic moieties and may, in fact, be a true informational RNA which can be incorporated by normal lymphocytes and then code for a specific immune response. In some situations, both mechanisms may be operative in varying degrees, but the data so far reported with RNA-mediated transfer of tumor immunity do not define which of these mechanisms is operative. In addition, it remains to identify the exact cell of origin of the immunogenic RNA, and it is not unreasonable to suspect that extraction of RNA selectively from this cell may produce RNA solutions of considerably greater efficacy.

REFERENCES

1. **Bard, D. S. and Pilch, Y. H.,** The role of the spleen in the immunity to a chemically-induced sarcoma in C3H mice, *Cancer Res.,* 29, 1125, 1969.
2. **Old, L. J., Boyse, E. A., Clarke, D. A., and Carswell, E. A.,** Antigenic properties of chemically induced tumors, *Ann. N.Y. Acad. Sci.,* 101, 80, 1962.
3. **Bard, D. S., Hammond, W. G., and Pilch, Y. H.,** The role of the regional lymph nodes in the immunity to a chemically-induced sarcoma in C3H mice, *Cancer Res.,* 29, 1379, 1969.
4. **Delorme, E. J. and Alexander, P.,** Treatment of primary fibrosarcoma in the rat with immune lymphocytes, *Lancet,* 2, 117, 1964.
5. **Wepsic, H. T., Zbar, B. M., Rapp, H. R., and Borsos, T.,** Systemic transfer of tumor immunity:delayed hypersensitivity and suppression of tumor growth, *J. Natl. Cancer Inst.,* 44, 955, 1970.
6. **Alexander, P., Delorme, E. J., and Hall, J. G.,** The effect of lymphoid cells from the lymph of specifically immunized sheep on the growth of primary sarcomata in rats, *Lancet,* 1, 1186, 1966.
7. **Balme, R. H., Dockerty, M. D., Grindlay, J. H., and Litzow, T. J.,** The use of immune lymph in the treatment of mouse cancer, *Minn. Med.,* 45, 892, 1962.
8. **Woodruff, M. F. A. and Nolan, B.,** Preliminary observations on treatment of advanced cancer by injection of allogeneic spleen cells, *Lancet,* 2, 426, 1963.
9. **Woodruff, M. F. A., Symes, M. O., and Stuart, A. E.,** The effect of rat spleen cells on two transplanted mouse tumours, *Br. J. Cancer,* 17, 320, 1963.
10. **Nadler, S. H. and Moore, G. E.,** Clinical immunologic study of malignant disease:response to tumor transplants and transfer of leucocytes, *Ann. Surg.,* 164, 482, 1966.
11. **Nadler, S. H. and Moore, G. E.,** Immunotherapy of malignant disease, *Arch. Surg.,* 99, 376, 1969.
12. **Fisher, B. and Fisher, E. R.,** Studies concerning the regional lymph node in cancer. I. Initiation of immunity, *Cancer,* 27, 1001, 1971.
13. **Fefer, A.,** Immunotherapy of primary maloney sarcoma-virus-induced tumors, *Int. J. Cancer,* 5, 327, 1970.
14. **Woodruff, M. F. A. and Boak, J. L.,** Inhibitory effect of pre-immunized CBA spleen cells on transplants of A strain mouse mammary carcinoma in (CBA × A)F_1 hybrid recipients, *Br. J. Cancer,* 19, 411, 1965.

15. **Borberg, H., Oettgen, H. F., Choudry, K., and Beattie, E. J., Jr.,** Inhibition of established transplants of chemically induced sarcomas in syngeneic mice by lymphocytes from immunized donors, *Int. J. Cancer,* 10, 539, 1972.
16. **Mannick, J. A. and Egdahl, R. H.,** Transfer of heightened immunity to skin homografts by lymphoid RNA, *J. Clin. Invest.,* 43, 2166, 1964.
17. **Ramming, K. P. and Pilch, Y. H.,** Mediation of immunity to tumor isografts in mice by heterologous ribonucleic acid, *Science,* 168, 492, 1970.
18. **Mannick, J. A. and Egdahl, R. H.,** Ribonucleic acid in "transformation" of lymphoid cells, *Science,* 137, 976, 1962.
19. **Ramming, K. P. and Pilch, Y. H.,** Transfer of tumor specific immunity with RNA: Inhibition of growth of murine tumor isografts, *J. Natl. Cancer Inst.,* 46, 735, 1971.
20. **Deckers, P. J. and Pilch, Y. H.,** Mediation of immunity to tumor-specific transplantation antigens by RNA: Inhibition of isograft growth in rats, *Cancer Res.,* 32, 839, 1972.
21. **Deckers, P. J. and Pilch, Y. H.,** RNA-mediated transfer of tumor immunity – a new model for the immunotherapy of cancer, *Cancer,* 28, 1219, 1971.
22. **Deckers, P. J., Edgerton, B. W., Thomas, B. S., and Pilch, Y. H.,** The adoptive transfer of concomitant immunity to murine tumor isografts with spleen cells from tumor-bearing animals, *Cancer Res.,* 31, 734, 1971.
23. **Hammond, W. G., Fisher, J. C., and Rolley, R. T.,** Tumor-specific transplantation immunity to spontaneous mouse tumors, *Surgery,* 62, 124, 1967.
24. **Prehn, R. T. and Main, J. M.,** Immunity to methylcholanthrene-induced sarcomas, *J. Natl. Cancer Inst.,* 18, 769, 1957.
25. **Deckers, P. J., Davis, R. C., Parker, G. H., and Mannick, J. A.,** The effect of tumor size on concomitant tumor immunity, *Cancer Res.,* 33, 33, 1973.
26. **Perper, T., Zee, T. W., and Mickelson, M.,** Purification of lymphocytes and platelets by gradient centrifugation, *J. Lab. Clin. Med.,* 72, 842, 1968.
27. **Cohen, A. M., Burdick, J. F., and Ketcham, A. S.,** Cell-mediated cytotoxicity: an assay using ^{125}I-iododeoxyuridine-labelled target cells, *J. Immunol.,* 107, 895, 1971.
28. **Attia, M. A. M. and Weiss, D. W.,** Immunology of spontaneous mammary carcinomas in mice. V. Acquired tumor resistance and enhancement in strain A mice infected with mammary tumor virus, *Cancer Res.,* 26, 1787, 1966.
29. **Klein, G.,** Tumor antigens, *Annu. Rev. Microbiol.,* 20, 223, 1966.
30. **Law, L. W.,** Studies of the significance of tumor antigens in the induction and repression of neoplastic disease, *Cancer Res.,* 29, 1, 1969.
31. **Old, L. J. and Boyce, E. A.,** Antigens of tumors and leukemias induced by viruses, *Fed. Proc.,* 24, 1009, 1965.
32. **Deckers, P. J., Edgerton, B. W., Thomas, B. S., and Pilch, Y. H.,** Tumor immunity in inbred mice with progressively growing syngeneic tumors. A study of concomitant immunity, *Curr. Top. Surg. Res.,* 3, 371, 1971.
33. **Rosenau, W. and Morton, D. L.,** Tumor-specific inhibition of growth of methylcholanthrene-induced sarcomas *in vivo* and *in vitro* by sensitized isologous lymphoid cells, *J. Natl. Cancer Inst.,* 36, 825, 1966.
34. **Barski, G. and Youn, J. K.,** Evolution of cell-mediated immunity in mice bearing an antigenic tumor. Influence of tumor growth and surgical removal, *J. Natl. Cancer Inst.,* 43, 111, 1969.
35. **Eilber, F. R. and Morton, D. L.,** Immunologic studies of human sarcomas: Additional evidence suggesting an associated sarcoma virus, *Cancer,* 26, 588, 1970.
36. **Gold, P., Gold, M., and Freidman, S. O.,** Cellular location of carcinoembryonic antigens of the human digestive system, *Cancer Res.,* 28, 1331, 1968.
37. **Hellström, I., Hellström, K. E., Pierce, G. E., and Yang, J. P. S.,** Cellular and humoral immunity to different types of human neoplasms, *Nature,* 220, 1352, 1968.
38. **Klein, G., Clifford, P., Klein, E., Smith, R. T., Minowada, J., Kourilsky, R. M., and Burchenal, J. H.,** Membrane immunofluorescence reactions of Burkitt's lymphoma cells from biopsy specimens and tissue cultures, *J. Natl. Cancer Inst.,* 39, 1027, 1967.
39. **Lewis, M. G., Ikonopisov, R. L., Nairn, R. C., Phillips, T. M., Fairley, G. H., Bodenham, D. C., and Alexander, P.,** Tumor-specific antibodies in human malignant melanoma and their relationship to the extent of the disease, *Br. Med. J.,* 3, 547, 1969.
40. **Morton, D. L., Malmgren, R. A., Holmes, E. C., and Ketcham, A. S.,** Demonstration of antibodies against human malignant melanoma by immunofluorescence, *Surgery,* 64, 233, 1968.
41. **Oettgen, H. F., Aoki, T., Old, L. J., Boyse, E. A., and DeHarven, E.,** Suspension culture of a pigment-producing cell line derived from a human malignant melanoma, *J. Natl. Cancer Inst.,* 41, 827, 1968.
42. **Hellström, I., Hellström, K. E., Pierce, G. E., and Bill, A. H.,** Demonstration of cell-bound and humoral immunity against neuroblastoma cells, *Proc. Natl. Acad. Sci. U.S.A.,* 60, 1231, 1968.
43. **Hellström, I., Hellström, K. E., Evans, C. A., Heppner, G. H., Pierce, G. E., and Yang, J. P. S.,** Serum-mediated protection of neoplastic cells from inhibition by lymphocytes immune to tumor-specific antigens, *Proc. Natl. Acad. Sci. U.S.A.,* 62, 362, 1969.
44. **Alexander, P., Delorme, E. J., Hamilton, L. D. G., and Hall, J. G.,** Effect of nucleic acids from immune lymphocytes on rat sarcomata, *Nature,* 213, 569, 1967.
45. **Deckers, P. J. and Pilch, Y. H.,** Transfer of immunity to tumor isografts by the systemic administration of xenogeneic immune RNA, *Nat. New Biol.,* 231, 181, 1971.

46. **Pilch, Y. H., Ramming, K. P., and Deckers, P. J.,** Transfer of tumor immunity with RNA, *Isr. J. Med. Sci.*, 7, 246, 1971.
47. **Kennedy, C. T. C., Cater, D. B., and Hartveit, F.,** Protection of C3H mice against BP-8 tumour by RNA extracted from lymph nodes and spleens of specifically sensitized mice, *Acta Pathol. Microbiol. Scand. (A)*, 77, 196, 1969.
48. **Pilch, Y. H., Fritze, D., Kern, D. H., and DeKernion, J. B.,** Immunotherapy of cancer with "immune" RNA, *Proc. Am. Assoc. Cancer Res.*, 16, 258, 1975.
49. **Lawrence, H. S.,** The transfer in humans of delayed skin sensitivity to streptococcal M. substance and to Tuberculin with disrupted leucocytes, *J. Clin. Invest.*, 34, 219, 1955.
50. **Lawrence, H. S. and Pappenheimer, A. M., Jr.,** Transfer of delayed hypersensitivity to diphtheria toxin in man, *J. Exp. Med.*, 104, 321, 1956.

IMMUNE RNA AND TUMOR IMMUNITY*

Yosef H. Pilch and David H. Kern

TABLE OF CONTENTS

*Supported, in part, by Grants CA 14846 and CA 15372 from the National Institutes of Health, Grant No. ET-28 from the American Cancer Society, and by grants from the California Institute for Cancer Research and the California Research Coordinating Committee of the University of California.

INTRODUCTION

In 1962 to 1964 Mannick and Egdahl first reported the mediation of immune responses to normal transplantation antigens with immune RNA (I-RNA). They transferred immunity to skin allografts in rabbits by incubating autologous spleen cells, in vitro, with RNA extracted from spleens and regional lymph nodes of specifically immunized rabbits. Reinfusion of such spleen cells incubated previously with I-RNA led to accelerated rejection of skin allografts taken from the same animal used as a source of tissue to immunize the RNA donor rabbit.[1-3] These findings were later confirmed by Sabbadini and Sehon[4-6] and in our own laboratory.[7]

It was also shown, a few years later, that immune responses to transplantation antigens in vitro could be mediated by I-RNA. Wilson and Wecker in 1966[8] and Bondevik and Mannick in 1968[9] reported that nonimmune lymphoid cells, incubated with I-RNA extracted from the lymphoid organs of animals immunized to histocompatibility antigens, caused specific in vitro immune cytolysis of target cells.

Since immune responses to normal transplantation antigens could be mediated by I-RNA, it seemed likely that immune responses to tumor antigens could also be mediated by I-RNA. In the 1960's, several authors working in a variety of systems published reports of the inhibition or retardation of tumor growth mediated by preparations of I-RNA. In 1967 Alexander et al.[10,11] reported growth retardation and occasional temporary regression of benzapyrene-induced sarcomas in rats treated by footpad injections of I-RNA extracted from lymph cells of sheep immunized with the particular rat tumor being treated. Rigby[12] induced prolonged survival of mice bearing Ehrlich ascites tumors by administering syngeneic spleen cells previously incubated with I-RNA from the spleens of mice immunized with this tumor. Londner et al.[13] reported a slowing of the growth of transplants of a spontaneously arising sarcoma in rats treated by the intraperitoneal injection of I-RNA extracted from the spleens of rats immunized with the tumor being treated. Kennedy et al.[14] described protection of C3H mice against transplants of a benzapyrene-induced sarcoma by footpad injections of I-RNA extracted from the lymph nodes and spleens of C57B1 mice immunized with this tumor. In this review, we will attempt to summarize the progress made in our laboratories over the past 6 years in the mediation of antitumor immune responses with preparations of I-RNA.

MEDIATION OF IMMUNITY TO MURINE TUMOR ISOGRAFTS BY XENOGENEIC IMMUNE RNA

In these studies xenogeneic antitumor I-RNA extracted from the lymph nodes and spleens of guinea pigs previously immunized to a chemically induced murine sarcoma was shown to transfer specific antitumor immune responses to mice of the strain in which the tumor was originally induced. This tumor-specific immunity was mediated by syngeneic spleen cells (obtained from nonimmune mice) incubated in vitro with antitumor I-RNA from guinea pigs and then administered intraperitoneally to previously untreated syngeneic mice. After injections of I-RNA, the mice were then challenged with inocula of viable tumor cells.[15-17]

Two benzapyrene-induced sarcomas (designated BP-4 and BP-1, induced and carried in female C3Hf/HeN mice) were used in their early transplant generations. Each of these tumors was found to possess distinct and different tumor-specific transplantation antigens, and no cross reactivity between the antigens of these two tumors could be found.[16] Groups of Hartley guinea pigs were immunized with each of these two tumors. I-RNA was extracted from the lymph nodes and spleens of these animals 10 to 14 days after immunization. Spleen cell suspensions were then prepared from the spleens of untreated C3H mice and incubated with the guinea pig I-RNA at 37°C for 20 min at a final RNA concentration of approximately 500 μg/ml (concentrations of 400 to 800 μg/ml proved to be equally effective). The concentration of spleen cells was then adjusted to 5 to 10 $\times$ 10^7 viable cells/ml. On two successive days, previously untreated C3H/HeN mice received intraperitoneal injections of 0.5 ml of these RNA-incubated spleen cell suspensions. On the first day all mice received a subcutaneous challenge inoculum of 10^4 BP-4 or 5 $\times$ 10^3 BP-1 viable tumor cells.

This model was then expanded to include a variety of additional experimental and control groups. In some experiments, the I-RNA solution was treated with ribonuclease, deoxyribonuclease, or pronase prior to incubation with spleen cells. In the course of these experiments, although not necessarily in every experiment, normal C3H/HeN spleen cells were incubated with the following experimental and control RNA preparations.

1. RNA from the lymphoid tissues of guinea pigs immunized with BP-4 mouse sarcoma.
2. RNA from guinea pigs immunized with BP-1 sarcoma.
3. RNA from guinea pigs immunized with normal C3H tissues.
4. RNA from guinea pigs injected with Freund's adjuvant only (Freund's adjuvant RNA).
5. BP-4 RNA treated with ribonuclease.
6. BP-4 RNA treated with deoxyribonuclease.
7. BP-4 RNA treated with pronase.

Normal C3H/HeN mice were divided into groups and were given, on two successive days, intraperitoneal injections of 0.5 ml containing 5 to 10 $\times 10^7$ spleen cells incubated previously with the I-RNA preparations designated for that group. Some control groups received normal C3H/HeN spleen cells which had not been incubated with I-RNA, and challenge control groups received no spleen cells at all. Mice in all groups received a subcutaneous inoculum of 10^4 BP-4 or 5×10^3 BP-1 tumor cells in the flank coincident with the first injection of spleen cells. All mice were then observed for subsequent tumor development.

The results of these experiments, which have been published in detail elsewhere,[15-17] demonstrated the following.

1. Inhibition in the development of specific isografts mediated by syngeneic spleen cells preincubated with xenogeneic antitumor I-RNA.
2. The abrogation of this response when the active BP-4 RNA preparation was treated with ribonuclease prior to incubation with spleen cells.
3. The failure of normal C3H tissue RNA to mediate this response.
4. The ineffectiveness of nonspecific guinea pig lymphoid RNA (Freund's adjuvant RNA).

Moreover, these responses were shown to be specific for the particular tumor used to immunize the guinea pigs from whose lymphoid organs the I-RNA was extracted. Note that while BP-4 RNA was effective in mediating immunity to isografts of BP-4, it was inactive against isografts of BP-1 and that BP-1 RNA transferred immunity to BP-1 isografts, but had no effect on isografts of BP-4. In addition, control experiments indicated that intraperitoneal injections of syngeneic spleen cells (without exposure) to RNA had no effect on the subsequent growth and development of tumor isografts.

These experiments indicated that immunity to tumor isografts in inbred mice could be mediated by syngeneic spleen cells preincubated with xenogeneic I-RNA extracted from the lymphoid organs of guinea pigs immunized with the specific tumor that was to be treated. Normal, nonimmune spleen cells appeared to be converted to specific immunoreactive status during this brief in vitro incubation with xenogeneic I-RNA. This immunity was manifested when such RNA-incubated spleen cells were injected into syngeneic mice, and the response was specific for the particular tumor used to immunize the guinea pig from whose lymphoid organs the I-RNA was prepared. Thus, spleen cells incubated with BP-4 RNA inhibited the growth of isografts of BP-4, but not isografts of the syngeneic but antigenically different tumor BP-1. Administration of spleen cells incubated with BP-1 RNA inhibited the growth of BP-1 isografts, but not BP-4 isografts. Nonspecific factors associated with guinea pig RNA probably did not account for the immunoreactivity acquired by normal mouse spleen cells following incubation with I-RNA, since RNA from guinea pigs immunized with Freund's adjuvant only was without effect. Likewise, injections of untreated syngeneic spleen cells (not incubated with RNA) did not affect the growth of tumor isografts. Mice treated in this manner had cumulative tumor incidences which did not differ significantly from those of control animals not infected with spleen cells. Most important was the finding that I-RNA from the lymphoid organs of guinea pigs immunized with normal C3H tissues failed to mediate an antitumor response. This, together with the finding that the effect of I-RNA for the immunizing tumor was specific, strongly suggested that the immunity expressed by the spleen cells preincubated with BP-4 RNA was directed primarily against tumor-associated antigens (TAA).

MEDIATION OF IMMUNITY TO RAT TUMOR ISOGRAFTS BY SYNGENEIC IMMUNE RNA

Experiments with xenogeneic I-RNA possessing

biologically active substances involved in tumor immunity provide convincing, but *indirect* evidence of the specificity of antitumor I-RNA for the tumor-associated antigens of the murine tumor under investigation. Because of certain obvious difficulties in the interpretation of results obtained in xenogeneic models and so that *direct* evidence that I-RNA mediates immune responses to TAA could be investigated, it was desirable to employ a system in which the tumor cells, the lymphoid tissues from which I-RNA was extracted, the spleen cells used for incubation with the I-RNA, and the recipient host all belonged to a single inbred strain. This would eliminate the possibility that the inhibition of tumor isograft development observed might be due, in part, to immune responses directed against transplantation or other membrane-associated antigens. In this model, a syngeneic tumor-host system in female Fischer 344/N rats was used. The tumor was a benz(a)pyrene-induced sarcoma, designed BP-1R, induced and carried in female Fischer 344/N rats. Syngeneic I-RNA was prepared from the spleens of Fischer rats after the excision of growing tumor transplants and was designated "BP-1R immune" RNA. Control RNA preparations designated "normal" RNA were prepared from the spleens of unimmunized Fischer 344/N rats never exposed to the BP-1R tumor.

Spleen cell suspensions were prepared from the spleens of untreated Fischer rats and were suspended in either "BP-1R immune" RNA or "normal" RNA at a concentration of 500 to 1,000 μg RNA/ml, and the concentration of spleen cells was adjusted to 5×10^7 to 10^8 spleen cells/ml. The RNA-spleen cell mixtures were incubated at 37° C for 20 min, washed once, and resuspended to a concentration of 5×10^7 to 5×10^8 viable "RNA-incubated" spleen cells/ml. Groups of female Fischer 344/N rats received 1 ml of I-RNA or normal RNA-incubated spleen cell suspensions intravenously via their tail veins daily for 3 days. Rats in all groups were inoculated subcutaneously with 1×10^3 viable BP-1R tumor cells in their left flank coincident with the first intravenous administration of "RNA-incubated" spleen cells. The second and third injections of "RNA-incubated" spleen cells were given 24 and 48 hr after the tumor isograft. Two additional control groups were included in these experiments. One group of rats received syngeneic spleen cells preincubated with "BP-1R immune" RNA treated previously with RNase.

The results of these experiments have been reported in detail elsewhere.[18] It was found that rats challenged with BP-1R tumor cells synchronous with the first of three daily intravenous injections of syngeneic spleen cells preincubated with syngeneic "BP-1R immune" RNA extracted from the spleens of rats immunized with the BP-1R sarcoma evidenced a much lower incidence of tumor isograft development than did rats in any other groups. The injection of syngeneic spleen cells alone, without incubation with RNA of any type, had no demonstrable antitumor effect. Furthermore, rats receiving spleen cells incubated with the same active "BP-1R immune" RNA preparation that was treated with RNase prior to incubation with the spleen cells manifested a tumor incidence similar to that of the challenge control animals, indicating that RNase treatment completely inactivated the "BP-1R immune" RNA. Moreover, "normal" syngeneic rat RNA was shown to be ineffective in mediating immune responses to the BP-1R tumor.

TRANSFER OF IMMUNITY TO TUMOR ISOGRAFTS BY SYSTEMIC ADMINISTRATION OF IMMUNE RNA

Earlier, in the mid 1960's, two investigators reported transferring immune responses by the direct systemic administration of I-RNA preparations. Friedman[19,20] transferred immunity to *Shigella*-soluble antigen in mice with intraperitoneal injections of syngeneic I-RNA prepared from the spleens of immunized mice. Alexander et al.[10,11] extracted xenogeneic I-RNA from the lymph cells of sheep immunized with chunks of primary benz(a)pyrene-induced rat sarcomas and injected these I-RNA preparations into the footpads of tumor-bearing rats. Animals receiving these I-RNA preparations exhibited a retardation in tumor growth and occasionally a complete regression of the tumors. It was apparent to us that a more direct and simpler method of utilizing I-RNA in mediating antitumor immune responses would involve the direct systemic administration of I-RNA itself, rather than the administration of lymphoid cells preincubated with I-RNA. In considering how to best achieve a transfer of immunity in vivo by the direct, systemic injection of I-RNA, our attention was drawn to the problem of the possible degradation of I-RNA by plasma and interstitial ribonucleases. Ribonuclease is known to inactivate preparations of I-RNA;

indeed, treatment of antitumor I-RNA preparations with RNase abolished antitumor immunoreactivity. Alexander et al. did not administer their xenogeneic I-RNA preparations intravenously because of the high levels of ribonuclease in blood. They injected I-RNA into the footpad. Moreover, these authors reported that the RNA preparations used in their successful experiments were not pure, containing considerable quantities of DNA as well as some protein – a "DNA-protein gel." No antitumor effect was noted with RNA extracts containing approximately 10% DNA and providing nonviscous solutions quite unlike the gellike suspensions obtained in their original method. They postulated that the presence of DNA, particularly in a gel form, might protect "biologically active" RNA from degradation by ribonucleases present in interstitial tissues.[10]

These facts led to the initial assumption that the direct systemic administration of xenogeneic I-RNA preparations (which were similar to the partially purified RNA extracts of Alexander) without prior incubation with lymphoid cells would be ineffective due to enzymatic degradation and inactivation of I-RNA by interstitial and plasma ribonucleases. We attempted to eliminate this problem by administering our I-RNA in a solution containing a potent RNase inhibitor.

Many polyanions are excellent inhibitors of RNase,[21-23] and the RNase inhibitor we selected was the polyanion sodium dextran sulfate. Previous studies in our laboratory had shown that sodium dextran sulfates of molecular weights 25,000 and 47,000 were poor inhibitors of ribonuclease, but that dextran sulfates of higher molecular weights (e.g., 77,500, 2.75×10^5, 5.0×10^5, and 2.0×10^6) were excellent ribonuclease inhibitors.[39] The BP-4 and BP-1 murine tumor-host systems described and the BP-1R tumor-host system in inbred Fischer 344/N rats described above were utilized for these studies.

Mediation of Immunity to Tumor Isografts in Rats

Xenogeneic I-RNA to the Fischer 344/N BP-1R rat tumor was prepared from lymphoid organs of Hartley guinea pigs immunized with BP-1R tumor cells in complete Freund's adjuvant. This RNA was designated I-RNA. A control RNA preparation was extracted from lymph nodes and spleens of unimmunized guinea pigs and designated "nonspecific" RNA. Sodium dextran sulfate (molecular weight 77,500) was added to each RNA preparation to a final concentration of 10 mg/ml. The RNA-dextran sulfate mixtures were then administered (as 0.25-ml doses injected into each footpad) to groups of normal Fischer rats every other day for 10 days. On the first day, rats in all groups received a subcutaneous challenge tumor cell inoculum of 10^4 viable BP-1R cells. Other control groups received tumor cells alone or tumor cells subcutaneously and footpad injections of dextran sulfate only (no RNA). Rats in all groups were observed biweekly for tumor development.

The results of this experiment, reported in greater detail elsewhere,[24,25] were as follows. Recipients of I-RNA from guinea pigs immunized with BP-1R administered in solutions containing sodium dextran sulfate exhibited significant resistance to the growth of isografts of BP-1R. Treatment with sodium dextran sulfate alone did not significantly alter the incidence of development of BP-1R isografts. This suggested that dextran sulfate did not, in and of itself, exert a nonspecific oncolytic or chemotherapeutic effect. Moreover, I-RNA from the nodes and spleens of unimmunized guinea pigs, administered in dextran sulfate, did not affect the growth of tumor isografts.

Mediation of Immunity to Tumor Isografts in Mice

Similar, but rather more involved, experiments were then performed in the BP-1 and BP-4 murine tumor-host systems. Xenogeneic I-RNA preparations to the BP-1 and BP-4 C3H/HeN mouse tumors were prepared from the lymph nodes and spleens of Hartley guinea pigs immunized with either BP-1 or BP-4 sarcoma cells in complete Freund's adjuvant. Control "normal tissue" RNA was extracted from the lymphoid organs of guinea pigs immunized with tissues from tumor-free C3H mice (lung, liver, kidney, and spleen) instead of tumor cells. Some I-RNA preparations were treated with RNase before use. RNA preparations used in these experiments were

1. RNA from the lymphoid organs of guinea pigs immunized with the BP-4 sarcoma (BP-4 RNA).
2. RNA from guinea pigs immunized with the BP-1 sarcoma (BP-1 RNA).
3. RNA from guinea pigs immunized with normal C3H tissues (C3H RNA).

4. BP-4 RNA treated with RNase (BP-4 RNA-RNase).

5. BP-1 RNA treated with RNase (BP-1 RNA-RNase).

Dextran sulfate (molecular weight 500,000) was then added to the RNA preparations to a final concentration of 3 mg/ml. The RNA-dextran sulfate mixtures were then administered (as 0.25-ml doses injected, subcutaneously, into each limb) to groups of normal C3H/HeN mice every other day for 6 days. One preparation of BP-4 RNA was administered without dextran sulfate. On the first day, all mice received a subcutaneous inoculum of 10^4 BP-4 or 5×10^3 BP-1 tumor cells.

Mice inoculated with BP-4 or BP-1 tumor cells and then treated with subcutaneous injections of tumor-specific xenogeneic I-RNA mixed with dextran sulfate demonstrated significant inhibition of tumor development when compared with mice which received tumor cells alone. Treatment of the active I-RNA preparations with RNase abolished this response. Specificity for the immunizing tumor was demonstrated in that the BP-1 RNA preparations induced immunity to BP-1 tumor isografts but failed to inhibit isografts to BP-4, whereas BP-4 RNA preparations induced immunity to BP-4 tumor isografts but failed to inhibit isografts to BP-1. Tumor specificity was also shown by the fact that normal tissue RNA (C3H-RNA) was ineffective. The administration of BP-4 RNA without dextran sulfate failed to inhibit the development of isografts of BP-4, indicating the importance of the RNase inhibitor for the activity, within our system, of systemically administered I-RNA preparations. When dextran sulfate alone was given, no suppression of tumor growth occurred, indicating that dextran sulfate itself did not exert a chemotherapeutic antitumor effect. Dextran sulfate presumably acts by preventing degradation of I-RNA by endogenous ribonucleases. In one experiment mice injected with BP-4 RNA with dextran sulfate were completely protected against the growth of BP-4 tumor isografts. The results of these studies have been previously published in detail.[24]

Although in most of our experiments the administration of I-RNA without dextran sulfate was ineffective, in one experiment we succeeded in transferring immunity to BP-4 tumor isografts with BP-4 RNA without dextran sulfate. Londner et al.[13] have reported the transfer of immunity to a rat sarcoma within a syngeneic system by intraperitoneal injections of syngeneic I-RNA, administering the RNA without an RNase inhibitor. Alexander et al.[10,11] successfully injected xenogeneic antitumor I-RNA directly into the footpads of rats without adding an RNase inhibitor. Kennedy et al.[14] employed a different RNase inhibitor. They prepared allogeneic I-RNA to a murine tumor and injected this RNA into the footpads of untreated mice in phosphate buffered saline pH 6.0 containing 0.1 mg/ml of polyvinyl sulfate (PVS), a potent RNase inhibitor. Mice so treated were rendered resistant to tumor isografts.

Immunotherapy of a Spontaneously Metastasizing Rat Mammary Carcinoma with Immune RNA

Recently, the successful immunotherapy of the transplantable spontaneously metastasizing mammary adenocarcinoma 13762 in female Fischer 344/N rats with xenogeneic or syngeneic I-RNA has been demonstrated. Growing primary tumor transplants surgically excised 18 days after transplantation do not recur locally. However, all animals die of metastases to lungs and other organs if no additional treatment is given. An experiment was performed in which immunotherapy with I-RNA was utilized as an adjuvant to surgical excision of the primary tumor transplant. The data from this experiment is illustrated in Figure 1. It can be seen that with no treatment at all, all animals died within 70 days. With excision of the primary tumor transplant alone, all animals died of metastases within 96 days. Xenogeneic I-RNA was extracted from the lymph nodes and spleens of Hartley guinea pigs 2 weeks following immunization with mammary adenocarcinoma tissue. This I-RNA was administered at a dose of 1.0 mg in 1 ml of buffer containing 10 mg/ml of sodium dextran sulfate (molecular weight 500,000), and 0.25 ml of this mixture was administered into each of the four footpads of a group of rats, every other day for ten doses, (i.e., 20 days) beginning 10 days prior to surgical excision of the primary tumor transplant. Four of five animals treated in this manner survived and remained free of disease for 180 days. Six of nine animals receiving xenogeneic I-RNA (at a dose of 1 mg every other day for ten doses beginning at the day of surgical excision of the primary tumor) survived and remained free of disease for more than 180 days. Syngeneic I-RNA was extracted from the spleens of Fischer rats

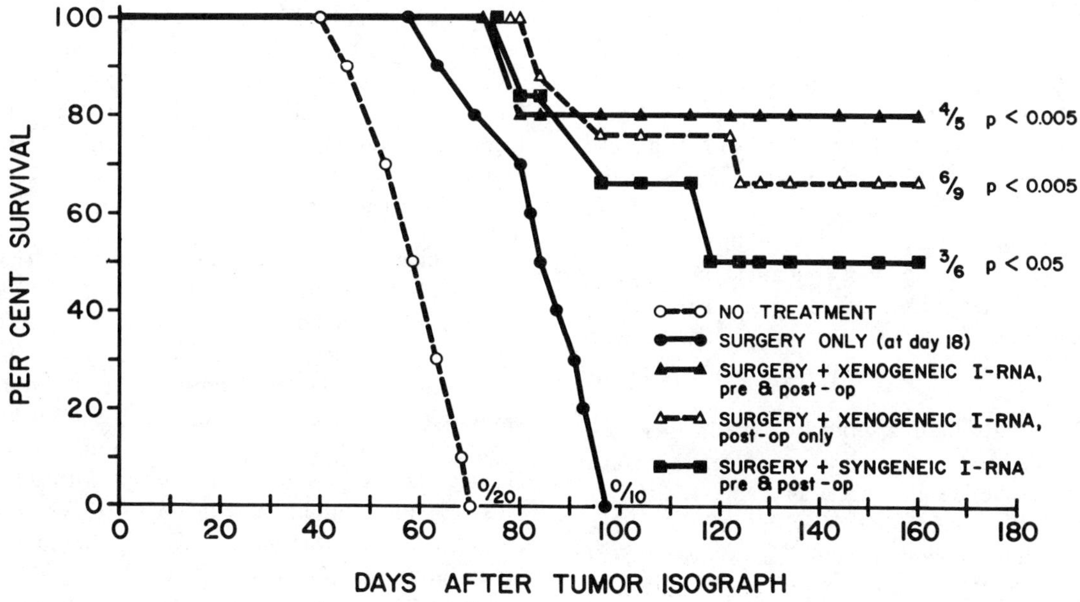

FIGURE 1. Immunotherapy of the spontaneously metastasizing mammary adenocarcinoma 13762 with xenogeneic and syngeneic I-RNA.

bearing growing transplants of the mammary adenocarcinoma. When this syngeneic I-RNA was administered, at the same dosage schedule, pre- and postoperatively to another group of rats, three of six animals so treated survived and remained free of disease for at least 180 days.

MEDIATION OF CYTOTOXIC ANTITUMOR IMMUNE RESPONSES, IN VITRO

Mediation of Cytotoxic Immune Reactions against Guinea Pig Sarcomas with Syngeneic Immune RNA

The semiquantitative "plaque" assay for cell-mediated immune cytolysis, first developed in our laboratory in 1968, was based on the technique of Granger and Weiser.[26] These studies utilized two sarcomas induced by 3-methylcholanthrene in inbred strain-2 guinea pigs. One, designated MCA-A, was a liposarcoma; the other, designated MCA-25, was an osteosarcoma. Each of these tumors was found, by Morton et al.[27] and by Holmes et al.,[28] to contain tumor-associated antigens and to be immunogenic in syngeneic animals.

Trypsinized cells from growing transplants of each tumor were grown as monolayers in plastic tissue culture flasks and serially subcultured. A line of normal strain-2 lung epithelial cells was also established in tissue culture and grown as monolayers in the same medium.

Syngeneic I-RNA for each tumor was obtained from strain-2 guinea pigs which were hyperimmunized with either tumor by excising growing tumor transplants followed by weekly injections of viable tumor cells. One week after the last immunization I-RNA was extracted from spleens and lymph nodes.

I-RNA from guinea pigs immunized with MCA-A was designated MCA-A RNA, and I-RNA from guinea pigs immunized with MCA-25 was designated MCA-25 RNA. In addition, for control purposes, I-RNA was extracted from the lymph nodes and spleens of strain-2 guinea pigs nonspecifically immunized with Freund's adjuvant. Some I-RNA preparations were treated with RNase prior to use. Spleen cells from normal, nonimmune strain-2 guinea pigs were incubated with one or another of the syngeneic I-RNA preparations for 20 min at 37°C and then washed. Aliquots of 0.25 ml of the washed RNA-incubated spleen cells containing 10^7 cells per 0.25 ml were dropped onto confluent monolayers of MCA-A, MCA-25, or normal lung epithelial cells. The cells dropped through the media and diffused into circular areas 1.5 to 2 cm in diameter on the tumor cell monolayers. The flasks were then placed in an incubator at 37°C. After 48 hr they were shaken, and the spleen cell droplets dislodged. The monolayers were washed once with tissue culture medium and inspected. Clear, acellular, macroscopic zones of lysis or "plaques" within the monolayer and corresponding to the areas covered by the spleen cells were considered evidence of immune cytolysis. Without immune cytolysis, the monolayers remained intact after elution of the spleen cell droplets. Droplets of normal strain-2 spleen cells not incubated with RNA were also applied to some monolayers as controls.

The results of this series of experiments have been reported previously.[29] However, some representative results appear in Table 1. Strain-2 spleen cells preincubated with MCA-A RNA induced zones of immune cytolysis in over 90% of the tests on MCA-A monolayers, but in only 23% of tests on MCA-25 monolayers. Likewise, spleen cells incubated in MCA-25 RNA produced areas of lysis in 16 of 18 tests on MCA-25 monolayers, but in only 3 of 15 tests on MCA-A monolayers. None of the I-RNA preparations mediated "plaque" formation on monolayers of normal strain-2 lung epithelial cells. Treatment of these syngeneic I-RNA preparations with RNase prior to incubation with spleen cells totally abolished the ability of these RNA-incubated spleen cells to produce immune cytolysis of target cells.

Mediation of Cytotoxic Immune Reactions against Murine Sarcomas with Xenogeneic Immune RNA

Later, more sophisticated microcytotoxicity tests were employed in a variety of animal and human tumor systems. Utilizing a quantitative assay for lymphocyte-mediated cytotoxicity, we reported in 1974 that xenogeneic I-RNA extracted from the lymphoid organs of guinea pigs immunized with mouse tumor cells mediated tumor-specific cytotoxic responses in vitro.[30]

Sarcomas were induced in female C3Hf/HeCr mice with either methylcholanthrene (MC-1) or 3,4-benz(a)pyrene (BP-1) and maintained by serial transplantation in female C3Hf/HeCr mice. The tumors were explanted from the second transplantation generation into plastic tissue culture flasks and grown as monolayers. Adult female Hartley guinea pigs were immunized with either MC-1 or BP-1 tumor cells in complete Freund's adjuvant. As controls, guinea pigs were immunized in a similar manner with a pool of normal C3H tissues. I-RNA was extracted from the lymphoid organs of the immunized guinea pigs 2 weeks following a single immunization. In some experiments, I-RNA was treated with RNase, DNase, or pronase prior to use. Spleen cells from normal nonimmune C3H mice were incubated with one or the other of the I-RNA preparations or with control RNA. The cytotoxicity of these spleen cells for either MC-1 or BP-1 tumor target cells following incubation with I-RNA was then measured in a microcytotoxicity assay in which tumor target cells were prelabeled with ^{125}I-iododeoxyuridine.[30]

In one experiment, representative of many performed in our laboratory, normal nonimmune

TABLE 1

Transfer of Tumor-specific Transplantation Immunity with RNA In Vitro

Groups	Number of plaques formed/ total number of tests: MCA-A monolayer	MCA-25 monolayer	Normal strain-2 lung monolayer
Strain-2 spleen cells incubated with RNA extracted from the lymphoid organs of strain 2 pigs			
Immunized with MCA-A	57/65	12/51[a]	0/15
Immunized with MCA-25	3/15	16/18[a]	0/14
Unimmunized	10/65	0/12	0/13

[a]$P < 0.01$ by chi-square (X^2) with Yates' correction when compared with identical cells dropped on MCA-A monolayers.

C3H mouse spleen cells preincubated with I-RNA extracted from the lymphoid organs of guinea pigs immunized with MC-1 tumor cells were found to be cytotoxic for MC-1 target cells, with a cytotoxicity index (CI) of 0.45 ± 0.04 (Figure 2). Spleen cells preincubated with RNA from guinea pigs immunized with either normal C3H tissues or with BP-1 tumor cells evidenced no significant cytotoxicity for MC-1 cells and were not statistically different from spleen cells incubated without RNA. Similar results were obtained at lymphocyte-to-target ratios of 100:1 and 50:1.

While 1 mg/ml of RNA was the concentration of I-RNA employed in the above experiment, we were interested in determining whether lower concentrations of I-RNA would be effective in mediating cytotoxic antitumor immune responses. Concentrations of RNA of 500, 200, and 100 μg/ml, as well as the usual 1 mg/ml, were studied. The results (Figure 3) indicated that a concentra-

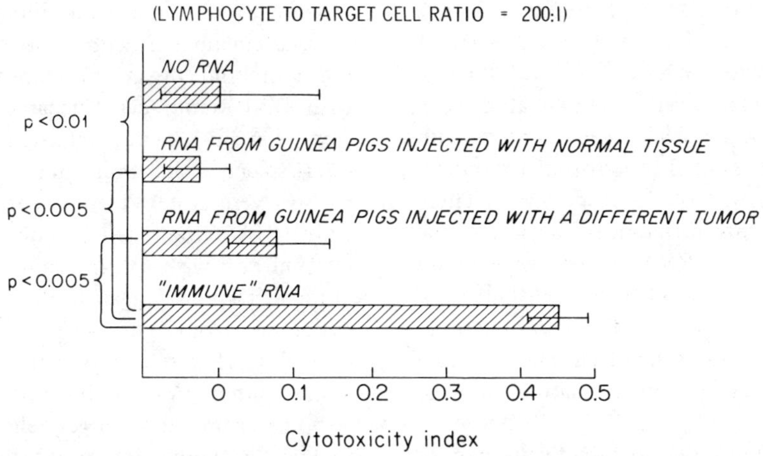

FIGURE 2. Immune cytolysis of MC-1 mouse tumor target cells mediated by xenogeneic I-RNA extracted from the lymphoid organs of guinea pigs immunized with MC-1 tumor cells. Controls include RNA from guinea pigs immunized with normal C3H mouse tissues or with antigenically distinct syngeneic tumor (BP-1).

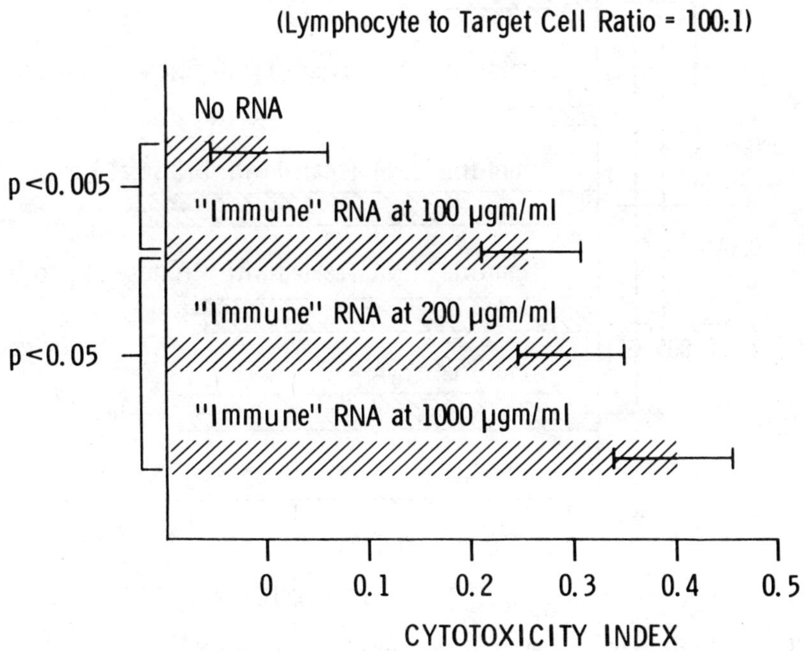

FIGURE 3. Immune cytolysis of MC-1 tumor cells mediated by xenogeneic I-RNA. Effects of varying I-RNA concentration.

tion of I-RNA as low as 100 μg/ml was sufficient to mediate significant cytolysis of MC-1 target cells.

Since the I-RNA preparations contained small but finite amounts of DNA and protein, the effects of RNase, DNase, and pronase treatment on the active anti-MC-1 I-RNA were evaluated. The results shown in Figure 4, of a single representative experiment, indicate that normal C3H spleen cells preincubated with anti-MC-1 I-RNA at a concentration of 1 mg/ml led to a cytotoxicity index of 0.32 ± 0.05. Anti-MC-1 I-RNA treated with RNase had no activity. However, C3H spleen cells preincubated with anti-MC-1 I-RNA that had been treated with DNAse or pronase resulted in cytotoxicity indices of 0.36 ± 0.06 and 0.28 ± 0.03, respectively. There were no significant differences between the activity of anti-MC-1 I-RNA treated with either DNase or pronase and untreated anti-MC-1 I-RNA. This suggests that the active component of these I-RNA preparations involved one or more species of RNA and not contaminating DNA or protein.

Utilizing BP-1 tumor cells as target cells, experiments similar to those described above were performed to study the specificity of the reactions mediated by these I-RNA preparations. The purpose of these experiments was to determine whether or not anti-MC-1 I-RNA could convert normal C3H spleen cells to cells cytotoxic to the antigenically distinct BP-1 tumor cells and whether or not anti-BP-1 I-RNA, which was not active in the MC-1 system, could mediate immune responses against BP-1 target cells. In these experiments, it was found that normal nonimmune C3H spleen cells, preincubated with xenogeneic I-RNA from the lymphoid organs of guinea pigs immunized with BP-1 tumor cells, became specifically cytotoxic for BP-1 target cells (Figure 5). However, C3H spleen cells preincubated with anti-MC-1 I-RNA were not cytotoxic for BP-1 target cells. Only RNA extracted from guinea pigs specifically immunized with BP-1 tumor cells mediated immune responses against BP-1 target cells. RNA extracted from guinea pigs immunized with normal C3H tissues (not shown) or with a different tumor (MC-1) did not mediate immune response against BP-1 target cells.

These studies indicated that normal non-

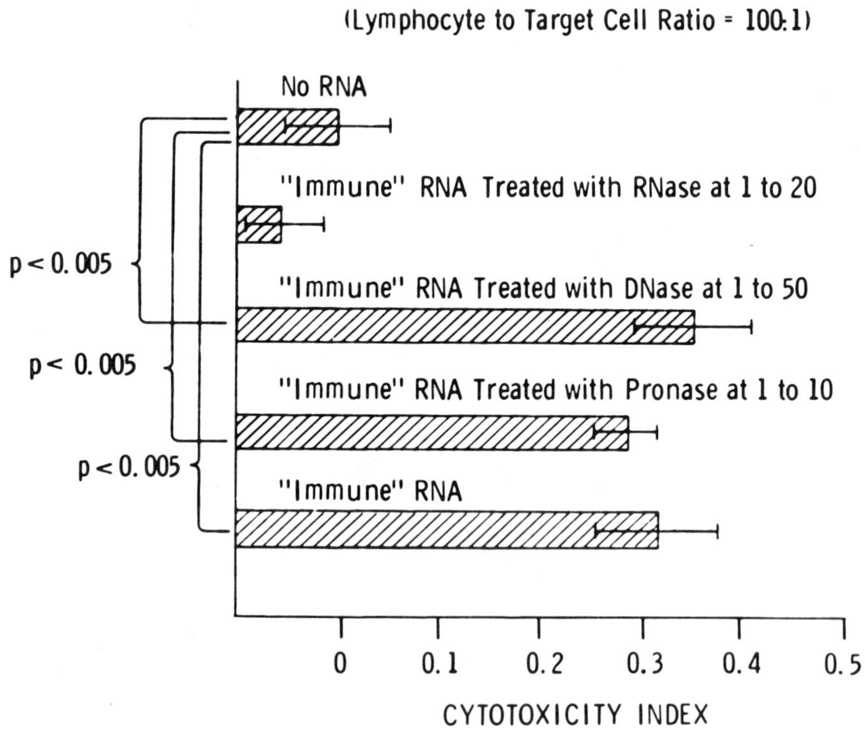

FIGURE 4. Effects of enzymes on the activity of anti-MC-1 I-RNA. RNase was used at an enzyme-to-substrate ratio of 1 to 20. DNase and pronase were used at enzyme-to-substrate ratios of 1 to 50 and 1 to 10, respectively, based on the actual amount of DNA or protein in the I-RNA preparation.

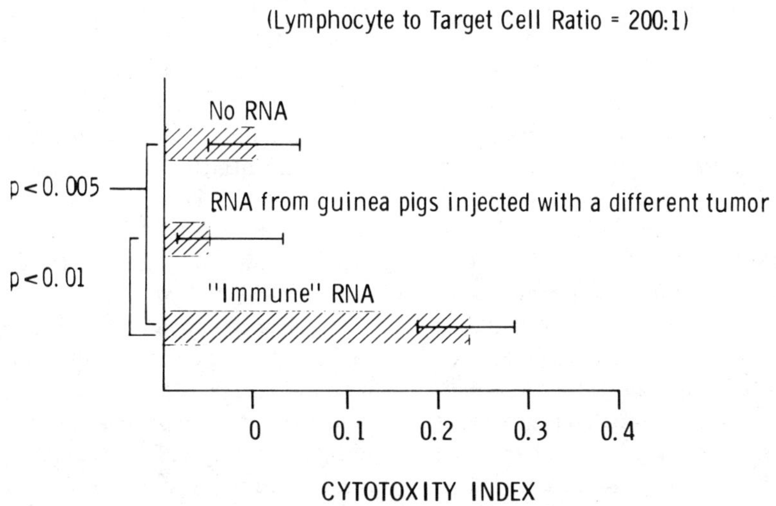

FIGURE 5. Immune cytolysis of BP-1 mouse tumor target cells mediated by xenogeneic I-RNA extracted from the lymphoid organs of guinea pigs immunized with BP-1 tumor cells. Control I-RNA was extracted from guinea pigs immunized with MC-1 tumor cells.

immune murine lymphoid cells could be converted to effector cells ("killer cells"), specifically cytotoxic to syngeneic murine tumor cells, by incubation, in vitro, with xenogeneic I-RNA extracted from the lymphoid organs of specifically sensitized guinea pigs. This response was shown to be specific for the particular tumor used to immunize the RNA donor animals. Normal mouse spleen cells incubated with I-RNA extracted from the lymphoid organs of guinea pigs immunized with MC-1 tumor cells were cytotoxic only to MC-1 target cells and not to target cells of the syngeneic but antigenically noncross-reactive tumor, BP-1. Spleen cells incubated with I-RNA from guinea pigs immunized with BP-1 tumor cells were cytotoxic to BP-1 target cells, but not to MC-1 target cells.

The guinea pigs immunized with mouse tumor cells were undoubtedly sensitized to a broad spectrum of murine transplantation antigens, as well as to the tumor-associated antigens present on the tumor cells. Yet, immunization with tumor cells appeared to be necessary for I-RNA to mediate an antitumor immune response, and these responses were specific for the particular tumor used to immunize the RNA donor. RNA extracted from the lymphoid organs of guinea pigs that had been immunized with normal C3H mouse tissues failed to induce C3H spleen cells to mediate immune responses against C3H tumor cells. It is likely that the anti-MC-1 I-RNA and anti-BP-1 I-RNA preparations we used contained I-RNA's directed against C3H transplantation antigens as well as to MC-1 or BP-1 tumor-associated antigens. Why, then, were the immune responses tumor specific? These I-RNA preparations were incubated with lymphoid cells which were syngeneic with respect to the tumor target cells. We postulate that, at least within our system, recognition of "self" occurred at the level of interaction between the I-RNA and the lymphoid cells. I-RNA's to C3H transplantation antigens were recognized as "self" by the C3H lymphoid cells, and, consequently, no immune response to C3H histocompatibility antigens was initiated. However, I-RNA's against tumor-associated antigens were recognized as "foreign" by C3H spleen cells, and specific antitumor immune responses to these antigens were elicited.

Mediation of Cytotoxic Immune Reactions against Rat Sarcomas with Syngeneic Immune RNA

Later, using a somewhat different quantitative cell-mediated microcytotoxicity assay (in which target cells are labeled with ^{3}H-thymidine rather than ^{125}I-iododeoxyuridine) in a rat tumor host system, we reported that syngeneic I-RNA directed specifically against the tumor-associated antigens of a chemically induced rat sarcoma mediated immune cytolysis of cells from the same rat sarcoma in vitro.[31] The tumor used the lymphoid tissues from which I-RNA was extracted, and the spleen cells incubated with the I-RNA were all obtained from members of a single inbred strain of rat. This was also the first report of the transfer of immunity to tumor-specific antigens with I-RNA extracted from lymphoid organs of tumor-bearing animals rather than hyperimmunized animals.

The model was a fibrosarcoma induced in a female Fischer 344/N rat by methycholanthrene and serially transplanted. The tumor, designated MC3-R, was explanted after the third transplant into plastic tissue culture flasks and grown as monolayers. I-RNA was prepared either from spleens of Fischer rats bearing 1- to 2-cm tumors or from spleens of rats hyperimmunized to this tumor by the method of Deckers and Pilch.[18] Control I-RNA preparations were also extracted from the spleens of Fischer rats bearing a different chemically induced syngeneic tumor, BP1-R (induced by benz(a)pyrene).

The results of these experiments indicated that normal nonimmune Fischer spleen cells preincubated with I-RNA extracted from the spleens of rats bearing growing MC3-R tumor transplants were significantly cytotoxic to MC3-R target cells (Figure 6). No significant cytotoxicity resulted using normal spleen cells incubated previously with I-RNA from the spleens of rats bearing BP1-R tumors. RNase treatment abolished the activity of the I-RNA preparations, but DNase or pronase treatment did not affect their activity (Figure 7). The concentration of RNA was apparently not critical within the limits tested (100 to 2,500 μg), although increasing RNA concentrations resulted in a moderate increase in the cytotoxic responses observed (Figure 8).

These experiments indicated, within a totally syngeneic system, that normal nonimmune lymphocytes could be converted to specifically cytotoxic effector cells ("killer" cells) by incubation, in vitro, with *syngeneic* I-RNA extracted from spleens of either tumor-bearing or hyperimmunized rats. The degree of cytotoxicity obtained following incubation with I-RNA was found to be comparable to the cytotoxic effects of spleen cells from rats hyperimmunized with the MC3-R tumor. Since the MC3-R and BP1-R tumors, the lymphoid tissues from which the I-RNA was extracted, and the spleen cells incubated with the I-RNA were all obtained from members of a single inbred strain, the immune responses elicited upon incubation of syngeneic spleen cells with syngeneic I-RNA very likely were directed only against tumor-associated antigens

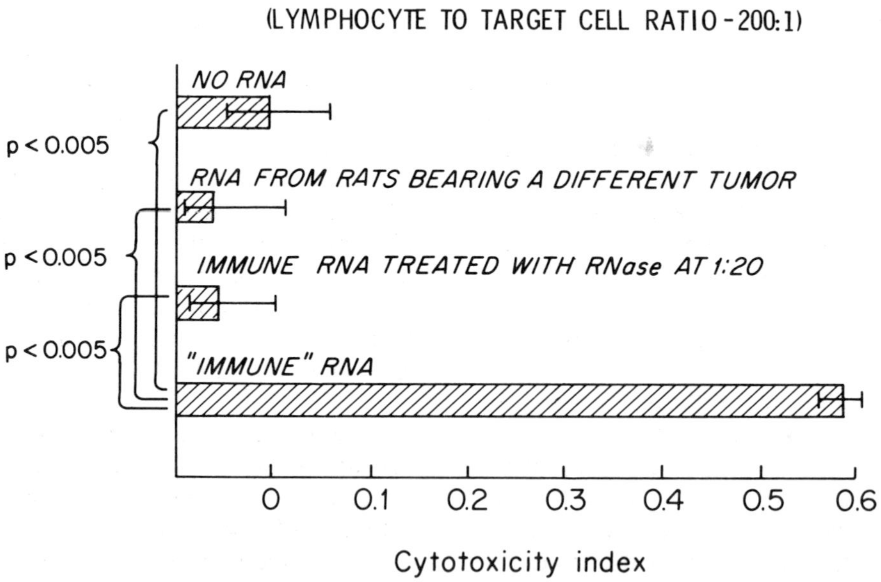

FIGURE 6. Immune cytolysis of MC3-R rat tumor cells mediated by syngeneic I-RNA extracted from the spleens of rats bearing growing MC3-R tumor transplants. Control RNAs included RNA from the spleens of rats bearing a different tumor (BP-1R) and active anti-MC3-R I-RNA which had been treated with RNase at an enzyme-to-substrate ratio of 1 to 20.

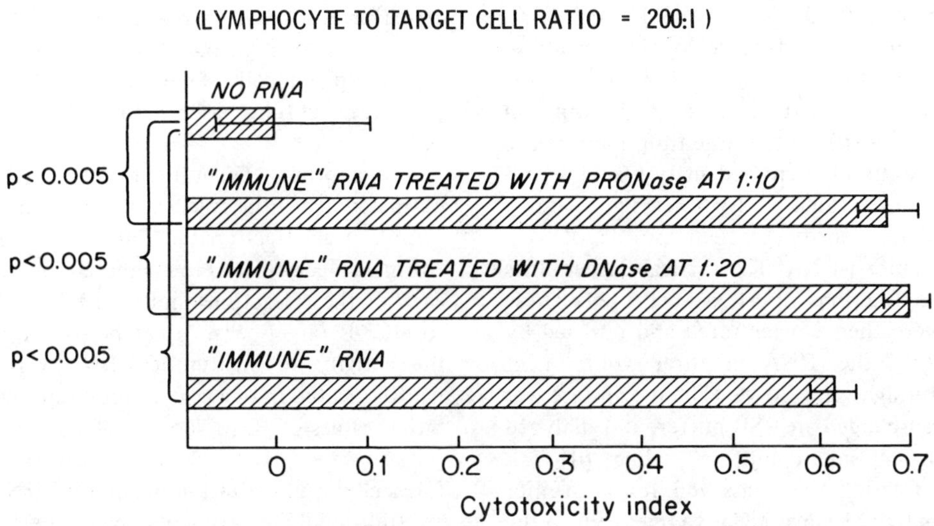

FIGURE 7. Effect of DNase and pronase treatment on the activity of anti-MC3-R I-RNA. The I-RNA was treated with DNase at an enzyme-to-substrate ratio of 1 to 20 and with pronase at an enzyme-to-substrate ratio of 1 to 10, based on the actual amount of DNA or protein in the I-RNA preparation.

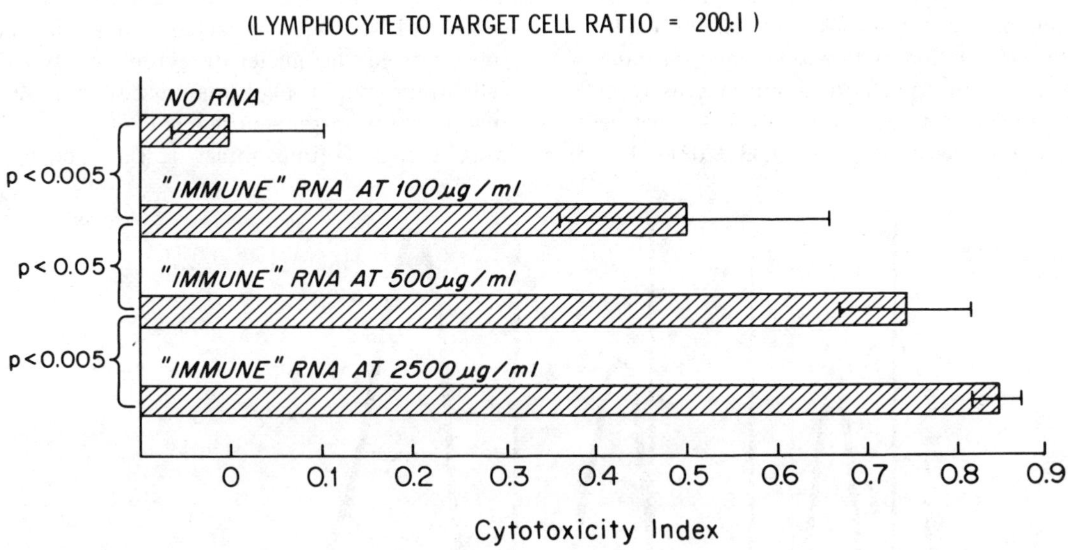

FIGURE 8. Effect of varying the I-RNA concentration on the immune cytolysis of MC3-R tumor cells.

expressed by the MC3-R tumor. They likely were mediated by an I-RNA specific for the tumor-associated antigens of the MC3-R sarcoma. This response was not due to nonspecific effects of rat lymphoid RNA, since spleen cells incubated with I-RNA from the spleens of rats bearing growing BP1-R tumors were not cytotoxic for MC3-R tumor cells. Again, these immune responses were abrogated when I-RNA was treated with RNase before incubation with spleen cells, but treatment with DNase or pronase did not effect the activity.

Fractionation of Syngeneic Immune RNA

In an attempt to identify the active component of the I-RNA preparations, antitumor I-RNA was fractionated by sedimentation velocity through sucrose density gradients. Syngeneic I-RNA extracted from spleens of Fischer rats bearing growing MC3-R tumor transplants was layered on top of 5 to 20% sucrose density gradients and centrifuged at 280,000 × g for 3 hr at 20° C. Fractions were collected by retrograde flotation. By monitoring the effluent at 254 nm, it was possible to collect fractions corresponding in sedimentation coefficients to the four major peaks of a high-resolution sucrose density gradient profile of the unfractionated I-RNA. Figure 9 is a typical sucrose density gradient profile of syngeneic antitumor I-RNA. The vertical dashed lines indicate the fractions of I-RNA collected. These fractions were then concentrated and purified by reprecipitating the RNA in three volumes of ethanol, overnight, at −20° C. The precipitate was washed, resuspended in RSB buffer, and dialyzed against cold phosphate buffered saline, pH 7.4. The RNA fractions were assayed for antitumor cytotoxic activity against MC3-R target cells in the microcytotoxicity assay previously described.[30] Aliquots of spleen cells from normal nonimmune Fischer rats were incubated separately with each of the four fractions of antitumor I-RNA isolated from the total RNA extracted from the spleens of Fischer rats bearing MC3-R tumors. I-RNA from the second fraction, with a sedimentation value of 8 to 16S, converted normal spleen cells to cells which were cytotoxic toward MC3-R target cells. Their cytotoxicity index was 0.33 ± 0.06 (Figure 10). However, RNAs with sedimentation values of 4 to 8S, 16 to 20S, or 20 to 35S, incubated with normal spleen cells, failed to lead to cells with significant cytolic effects toward MC3-R target cells.

In another experiment, syngeneic I-RNA extracted from the spleens of rats bearing MC3-R tumors was fractionated into six components. Normal spleen cells were incubated separately with each of these fractions and tested for cytotoxicity to MC3-R target cells. As can be seen in Figure 11, the activity of antitumor I-RNA was predominantly located in one region, representing sedimentation values of 12 to 16S.

Intracellular Localization of Immune RNA

Additional experiments were performed to determine the intracellular localization of immunologically active immune RNA.[32] Syngeneic I-RNA was extracted from intact spleens from the nuclei or cytoplasm of spleen cells from rats bearing growing MC3-R tumor transplants. Xenogeneic I-RNA was extracted from the intact spleens from the nuclei or cytoplasm of spleen cells from guinea pigs immunized with MC3-R tumor cells. Control xenogeneic RNA preparations were prepared from intact spleens, nuclei, or

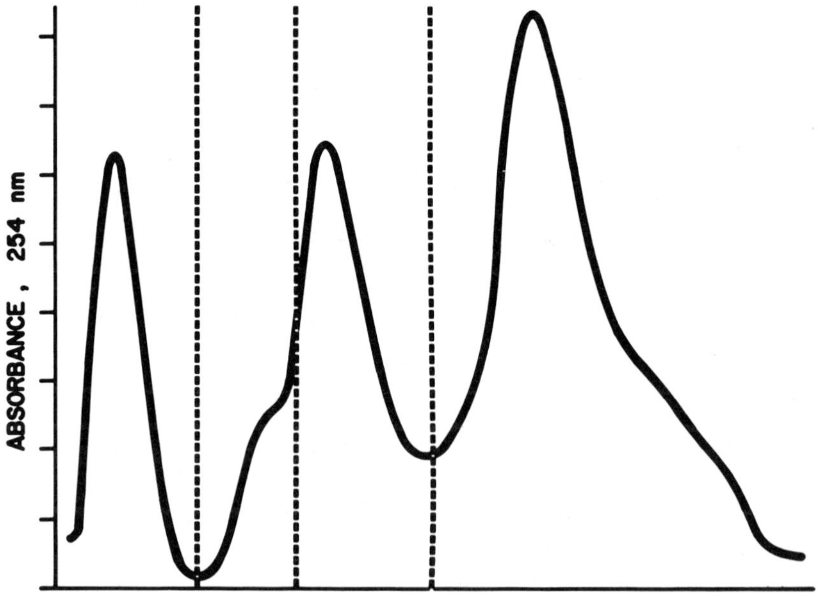

FIGURE 9. A sucrose density gradient profile of syngeneic I-RNA extracted from the spleens of rats bearing growing MC3-R tumor transplants. The low molecular weight (4S) RNA is to the left, and the higher molecular weight (28S) RNA is to the right. Vertical dashed lines indicate the points at which the RNA was separated into four fractions.

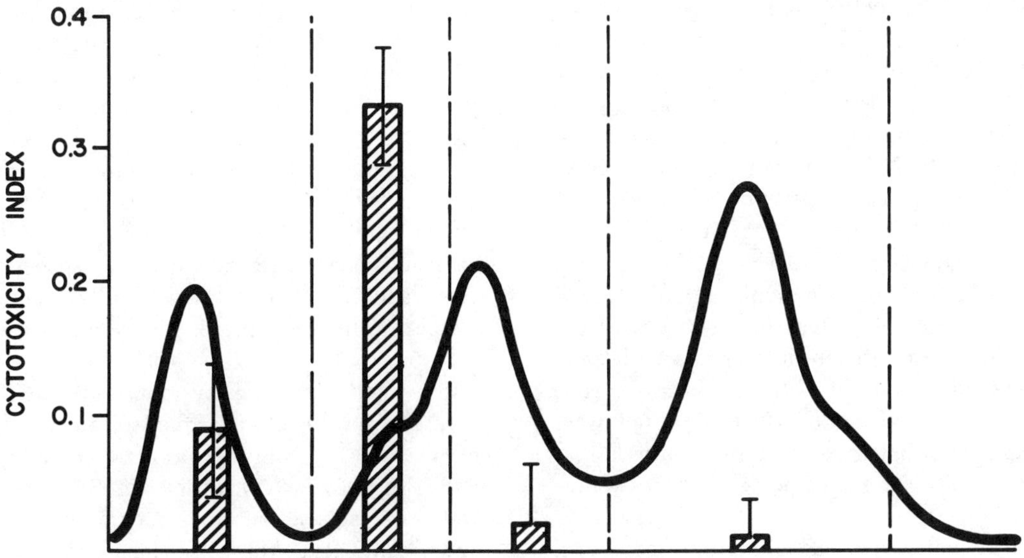

FIGURE 10. Localization of the active moiety of syngeneic I-RNA extracted from the spleens of rats bearing growing MC3-R tumors. Aliquots of normal rat spleen cells were incubated with one of the four fractions of anti-MC3-R I-RNA, and the cytotoxic effects of these RNA-incubated spleen cells on MC3-R target cells were measured.

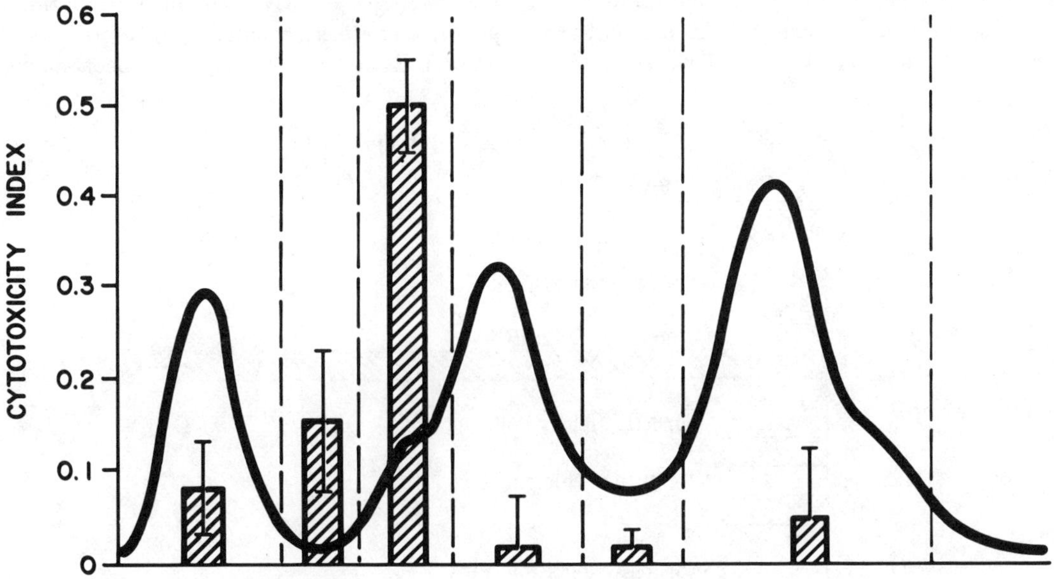

FIGURE 11. Localization of the active moiety of syngeneic I-RNA extracted from the spleens of rats bearing growing MC3-R tumors. Aliquots of normal rat spleen cells were incubated with one of the six fractions of anti-MC3-R I-RNA, and the cytotoxic effects of the cells on MC3-R target cells were then measured.

cytoplasm of spleen cells of guinea pigs immunized with complete Freund's adjuvant from intact spleens, nuclei, or cytoplasm of spleen cells from normal Fischer rats. Allogeneic spleen cells from normal nonimmune Wistar-Furth rats or syngeneic spleen cells from normal nonimmune Fischer rats were incubated separately with each of these various I-RNA preparations, and the cytotoxic effects of these cells for MC3-R target cells were measured in a cell-mediated microcytotoxicity assay.[30]

The cytotoxic effects of allogeneic effector

cells following incubation with xenogeneic I-RNA fractions toward MC3-R tumor cells are depicted in Figure 12. Normal Wistar-Furth rat spleen cells, incubated with xenogeneic I-RNA extracted from the cytoplasmic fraction of spleens of guinea pigs immunized with MC3-R tumor cells, became as cytotoxic toward MC3-R target cells as spleen cells incubated with I-RNA from the intact spleens of immunized guinea pigs. Nuclear RNA from the spleens of guinea pigs immunized with MC3-R tumor cells was completely inactive. RNA from guinea pigs nonspecifically immunized with complete Freund's adjuvant was also inactive. It is likely that the cytotoxic effects detected were directed, at least in part, against histocompatibility differences between guinea pigs and MC3-R tumor cells.

Previous work from our laboratory had shown that after incubation of xenogeneic I-RNA with normal spleen cells syngeneic with respect to the immunizing tumor, the resultant cytotoxic immune responses (as determined on target cells of the same tumor source) are directed specifically against tumor-associated antigens of that tumor.[30] Therefore, aliquots of the same RNAs used in the previous experiment were incubated with normal nonimmune syngeneic Fischer rat spleen cells, and the resultant cytotoxicity for MC3-R target cells was measured. I-RNA extracted either from whole spleen cells or from the cytoplasmic fraction of spleen cells from guinea pigs immunized with MC3-R tumor cells converted normal Fischer rat spleen cells to cells cytotoxic toward MC3-R target cells (Figure 13). RNA from the nuclear fraction was inactive. RNA extracts from the spleen or the cytoplasmic or nuclear fractions of spleens from nonspecifically immunized guinea pigs were also inactive.

In a totally syngeneic system, syngeneic I-RNA prepared from the cytoplasm of spleen cells of rats bearing MC3-R tumors was as active as RNA from whole spleens in converting normal syngeneic Fischer rat spleen cells to react immunologically toward MC3-R target cells (Figure 14). In this syngeneic system histocompatibility differences were illuminated, and these immune responses very likely were directed against the tumor-associated antigens of the MC3-R target cells.[31] The immunologic activity was confined mainly to the fractions of RNA obtained from the cytoplasm; a relatively small amount of activity of the nuclear RNA fraction may have been due to the

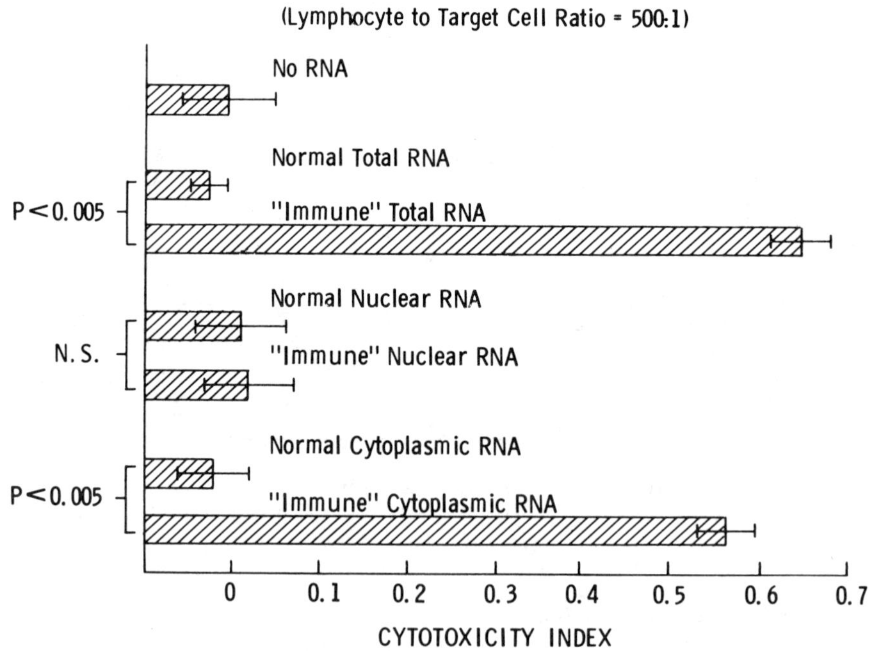

FIGURE 12. Immune cytolysis of MC3-R tumor target cells mediated by xenogeneic I-RNA extracted from intact spleen cells and from nuclear and cytoplasmic fractions of spleen cells from guinea pigs immunized with MC3-R tumor cells. The effector cells were spleen cells from normal allogeneic Wistar-Furth rats.

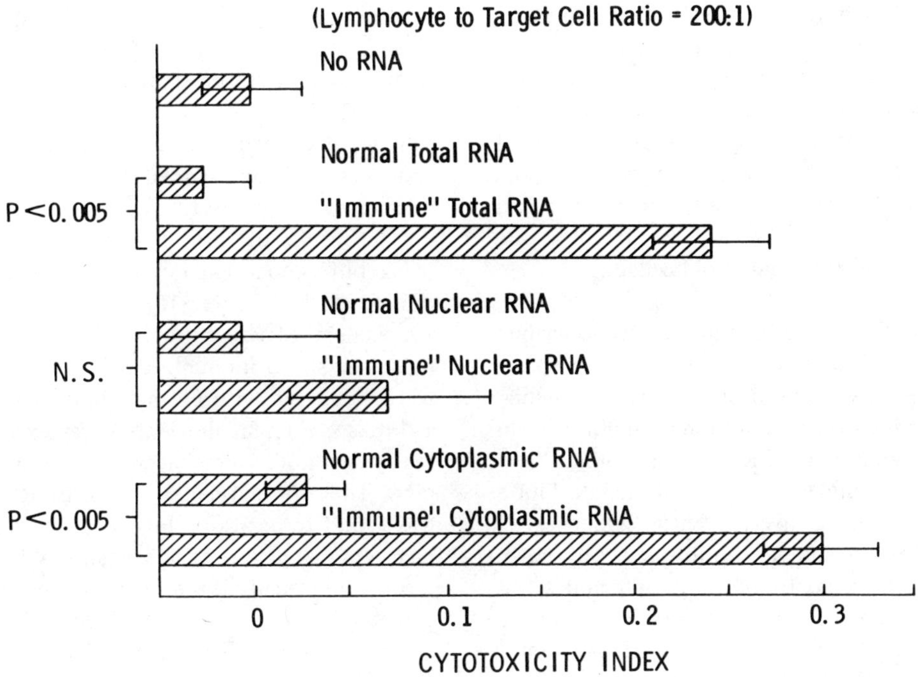

FIGURE 13. Immune cytolysis of MC3-R tumor target cells mediated by xenogeneic I-RNA extracted from intact spleen cells and from nuclear and cytoplasmic fractions of spleen cells from guinea pigs immunized with MC3-R tumor cells. The effector cells were spleen cells from normal syngeneic Fischer rats.

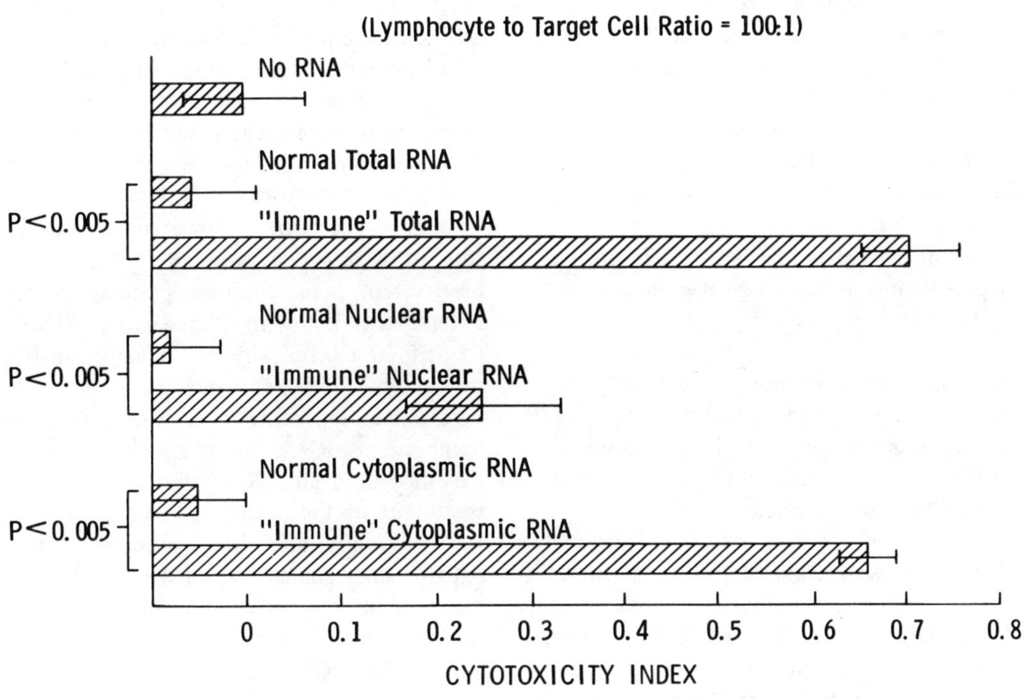

FIGURE 14. Immune cytolysis of MC3-R tumor target cells mediated by syngeneic I-RNA extracted from intact spleen cells and from nuclear and cytoplasmic fractions of spleen cells from Fischer rats bearing growing MC3-R tumor transplants. The effector cells were spleen cells from normal syngeneic Fischer rats.

presence of a significant amount of cytoplasmic contamination of the nuclei preparations.

The cytoplasmic fractions of both syngeneic and xenogeneic I-RNa were as active on a weight-per-weight basis as their respective unfractionated I-RNA counterparts. These data suggest that the immunologically active fraction of syngeneic and xenogeneic antitumor I-RNA is confined to the cytoplasm of sensitized lymphoid cells.

Mediation of Cytotoxic Immune Reactions against Human Gastric Carcinoma Cells

In studies conducted in our laboratory, mediation of immune cytolysis of human tumor cells, in vitro, by xenogeneic I-RNA was found.[33-36] Normal nonimmune human peripheral blood lymphocytes were converted into "killer cells" cytotoxic toward human tumor cells, in vitro. I-RNA was extracted from the lymph nodes and spleens of sheep or guinea pigs previously immunized with cells from the particular human tumor being studied. The human tumors employed were

1. A human gastric adenocarcinoma, RI-H, obtained by primary explant from a surgical specimen established in tissue culture in our laboratory and studied in its 12th to 14th passage.
2. A human adenocarcinoma of the breast, BT-20, obtained from Dr. E. Lasfargues and studied in its 300th to 310th passage.
3. A human malignant melanoma, ED-H, explanted from a surgical specimen in the laboratory of Dr. Donald L. Morton, UCLA, and studied in its 18th to 25th passage.
4. A human melanoma, RO-H, established by primary explant from a surgical specimen and studied in its 10th to 15th passage.

Sheep or guinea pigs were immunized with viable tumor cells from each of these tissue culture lines in complete Freund's adjuvant. In some experiments I-RNA was treated with ribonuclease, deoxyribonuclease, or pronase.

Peripheral blood leukocytes from normal human volunteers were isolated from heparinized specimens of peripheral blood by sedimentation in 10% Plasmagel (Roger Bellon Laboratories, Neuilly, France).[33] In subsequent experiments, purified mononuclear cell populations were prepared from heparinized blood specimens on Ficol-Isopaque radients, suspended in RPMI 1640 with 40% agamma human serum and 10% dimethysulfoxide (DMSO), frozen in liquid nitrogen at 1° C/min, and thawed at 37° C prior to use.[33] These lymphocytes were incubated for 20 min at 37° C with I-RNA at a concentration of 2 mg/ml. The cells were washed, diluted to the appropriate cell concentration, and added to target cells in the microcytotoxicity assay.[30,33]

The results of three typical early experiments using buffy coat leukocytes as effector cells are presented in Figure 15. In two experiments, xenogeneic I-RNA extracted from the lymphoid organs of sheep immunized with RI-H tumor cells were incubated with normal human leukocytes. In a third experiment the I-RNA was extracted from the lymph nodes and spleens of immunized guinea pigs. The optimal leukocyte-to-tumor cell ratio was found to be 500:1; higher ratios resulted in an unacceptable degree of nonspecific lysis, and lower ratios were usually less effective or ineffective. The cytotoxicity index (CI) obtained with leukocytes incubated with I-RNA extracted from the lymphoid organs of immunized sheep (sheep 1) was 0.45 ± 0.08. Leukocytes incubated with RNA extracted from the lymphoid organs of a sheep immunized with complete Freund's adjuvant only (Freund's adjuvant RNA) gave a CI value significantly different ($p < 0.005$) from I-RNA, but not different from the leukocyte preparation incubated without RNA. The response was almost totally abrogated after aliquots of the I-RNA preparation were treated with ribonuclease prior to incubation with leukocytes. (This preparation of RNase was subsequently found to be partially inactive. As described below, when a fresh preparation of RNase was used, the I-RNA was totally inactivated. In a second experiment using sheep I-RNA, leukocytes incubated with I-RNA gave a CI of 0.30 ± 0.05. This was a significantly greater cytotoxicity index than was obtained with leukocytes incubated without RNA or leukocytes incubated with RNA from animals injected with Freund's adjuvant. Note that in this experiment treatment of the active I-RNA with RNase prior to use totally abolished its activity. In a third experiment, guinea pig I-RNA was employed instead of sheep I-RNA, but all other factors were similar. The CI for leukocytes incubated with guinea pig I-RNA was 0.41 ± 0.13. This was a significantly higher cytotoxicity index than was obtained with leukocytes incubated without RNA, leukocytes incubated with RNA from animals immunized with Freund's adjuvant, and leukocytes incubated with I-RNA treated with ribonuclease.

The use of mononuclear cell populations, prepared on Ficol-Isopaque gradients instead of unfractionated buffy coat leukocytes incubated with the I-RNA, led to an increase in the cytotoxicity index, even with a reduction in the lymphocyte-to-target cell ratio (from 500:1 to 100:1 or 200:1, Figures 16 and 17). At a lymphocyte-to-target cell ratio of 100:1, a cytotoxicity index of 0.66 ± 0.08 was obtained (Figure 16).

Since lymphocytes normally account for 20 to 40% of the total peripheral white blood cells in man, this result is quite consistent with what would be expected assuming that one or more population of lymphocytes is the active effector cell population. By effecting approximately a fivefold increase in lymphocyte concentration, the effector cell-to-target cell ratio could be reduced by a factor of 5, with no loss (or even an increase) in the cytotoxic effects. Pretreatment of active I-RNA preparations with deoxyribonuclease (Figure 17) or with pronase[33] (not shown) did not alter the cytotoxicity index obtained. The fact that deoxyribonuclease and pronase treatment of the I-RNA preparations did not affect the biological activity of the RNA extracts, but that ribonuclease treatment abrogated the response (Figure 16), indicated that RNA was likely to be involved in the immunological activity of the preparations.

Next, the effects of varying the RNA concentration upon the magnitude of the immune response obtained were determined. The concentration of RNA used for incubation with lymphocytes was progressively reduced so that a dose-response curve might be obtained and a "threshold" RNA concentration determined. The results of one experiment are depicted in Figure 18. From these data it can be seen that reducing the RNA concentration from 2,000 to 1,000 μg/ml or even to 500 μg/ml had little effect on the response. Reducing the concentration still further, to 250 μg/ml, resulted in approximately a 50% decrease in cytotoxicity (not shown). In other experiments it was found that reducing the RNA

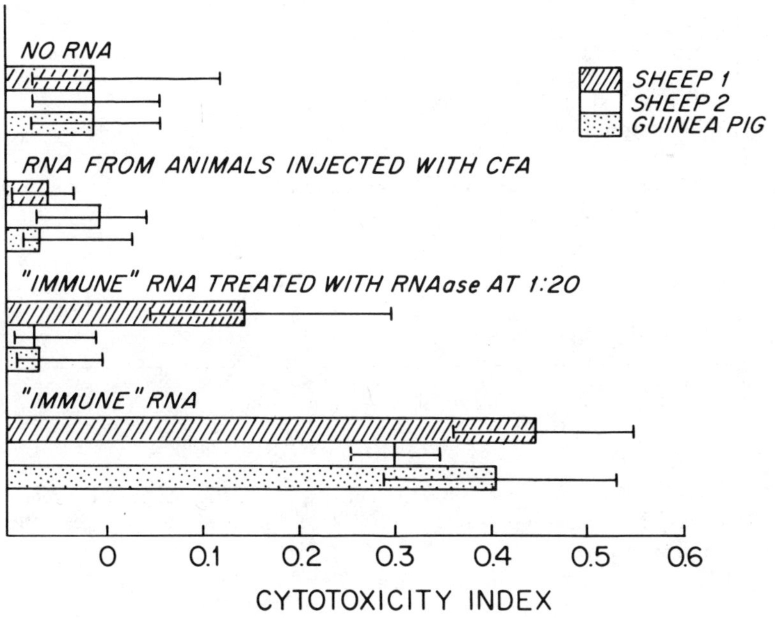

FIGURE 15. Immune cytolysis of RI-H gastric carcinoma cells mediated by xenogeneic I-RNA extracted from lymphoid organs of sheep and guinea pigs following immunization with RI-H cells. Experiments 1 and 2 utilized sheep I-RNA, and experiment 3 employed guinea pig I-RNA. Differences between I-RNA and all control groups were statistically significant ($p<0.005$ to $p<0.05$) in every case. RNase-treated I-RNA in experiment 1 was only partially degraded, but a different lot of RNase produced complete degradation in experiments 2 and 3. Ratio of leukocytes to target cells was 500:1. CFA= complete Freund's adjuvant.

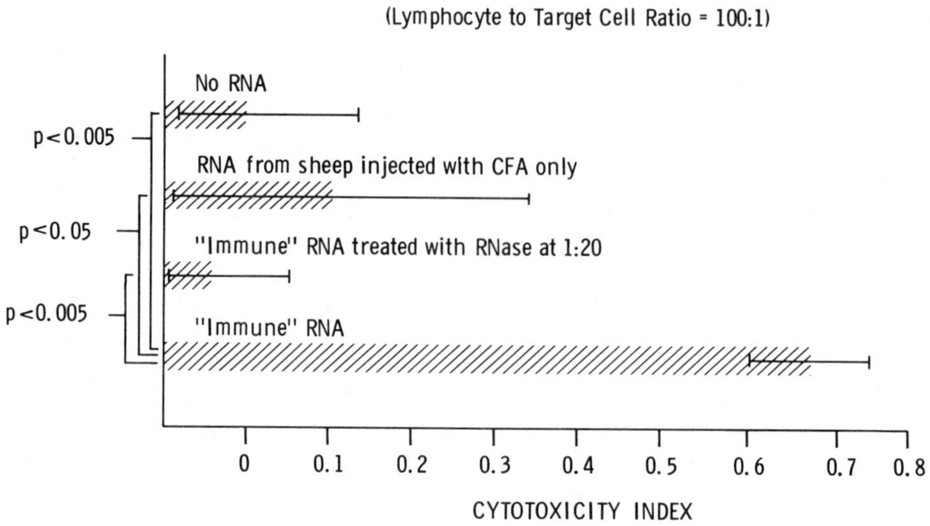

FIGURE 16. Immune cytolysis of RI-H tumor cells mediated by xenogeneic I-RNA extracted from the lymphoid organs of sheep immunized with RI-H tumor cells. The mononuclear effector cells were purified from whole blood on Ficoll-Isopaque gradients. Controls include RNA from sheep injected with complete Freund's adjuvant (CFA) only and active I-RNA after treatment with RNase at an enzyme-to-substrate ratio of 1 to 20.

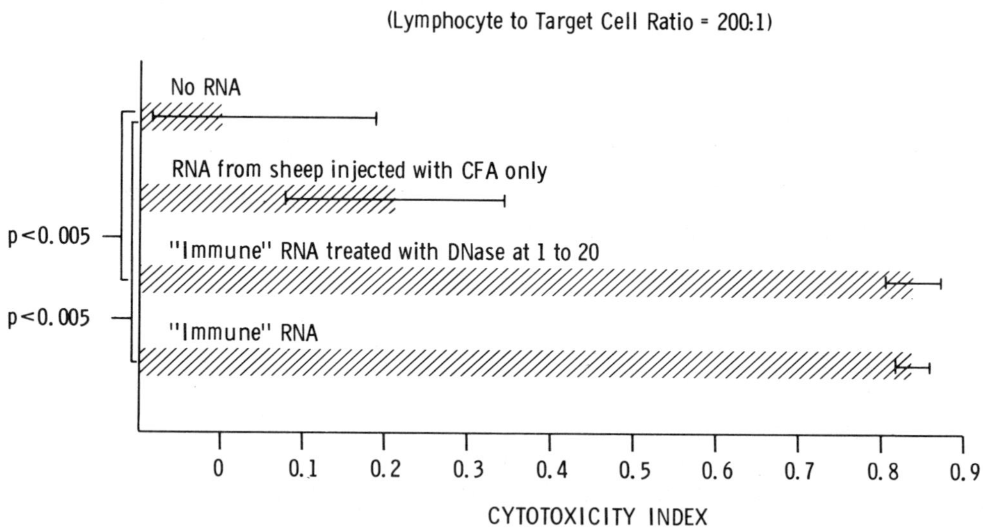

FIGURE 17. Effect of DNase treatment on the activity of I-RNA. Active xenogeneic I-RNA was treated with DNase at an enzyme-to-substrate ratio of 1 to 20, based on the actual amount of DNA present in the RNA preparations.

concentration to 100 μg/ml reduced cytotoxicity still further, but this result remained significantly greater than that obtained without RNA.

Mediation of Cytotoxic Immune Reactions against Human Breast Carcinoma Cells

Experiments similar to those described above for the RI-H cell line were performed utilizing the BT-20 breast carcinoma cell line. In these experiments, xenogeneic I-RNA from immunized guinea pigs was used instead of sheep I-RNA. The results are depicted in Figure 19. Xenogeneic I-RNA extracted from the lymphoid organs of guinea pigs immunized with BT-20 tumor cells induced normal nonimmune human lymphocytes to become cytotoxic toward BT-20 tumor target

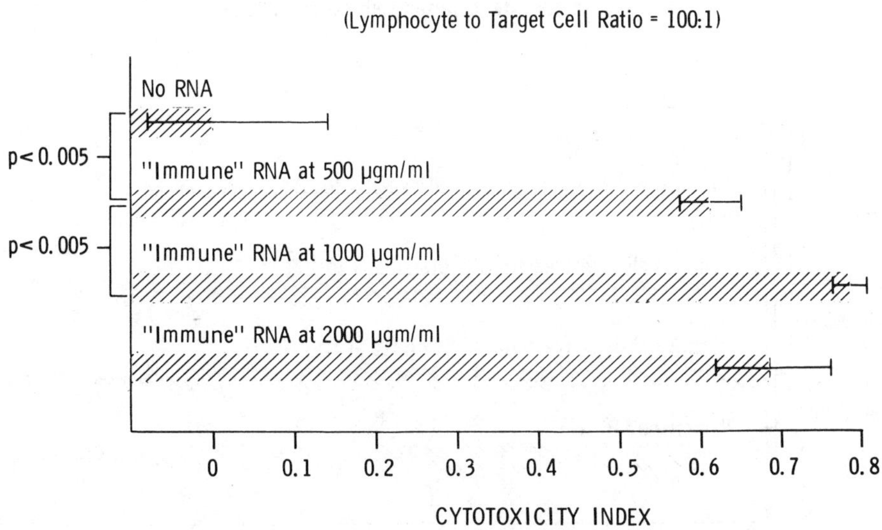

FIGURE 18. Effect of varying the concentration of I-RNA on the cytolysis of RI-H cells.

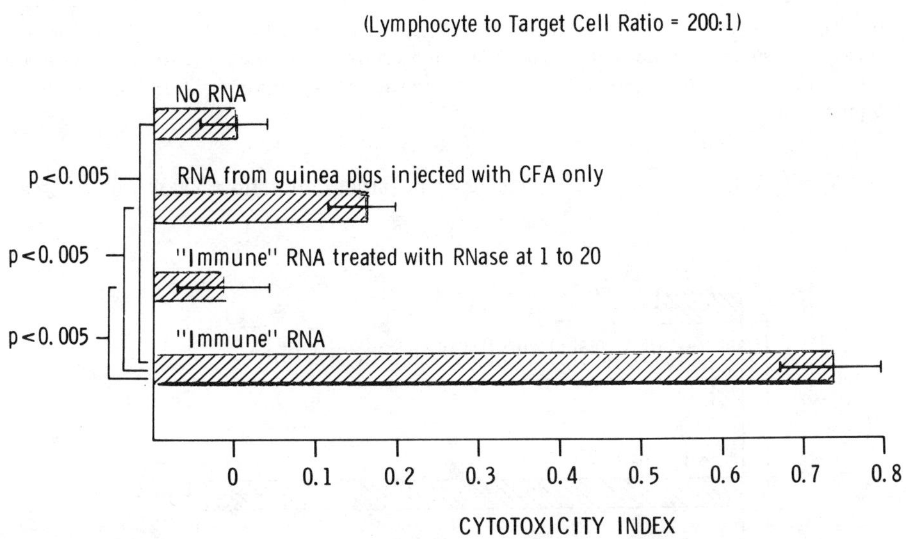

FIGURE 19. Immune cytolysis of BT-20 breast cancer cells mediated by xenogeneic I-RNA extracted from the lymphoid organs of guinea pigs immunized with BT-20 tumor cells.

cells. RNA from guinea pigs immunized with Freund's adjuvant only was inactive (Figure 20). The activity of the I-RNA was abolished by treatment of the I-RNA preparations with RNase, but was not affected by treatment with DNase or pronase.

Mediation of Cytotoxic Immune Reactions Specifically Directed against Tumor-associated Antigens of Human Cancer Cells

Experiments were then designed to extend our studies to a third histologic type of human cancer, malignant melanoma. Melanoma cell lines ED-H and RO-H were previously established in tissue culture from primary explants of surgical specimens obtained from melanoma patients ED and RO, respectively. In the initial experiments I-RNA was extracted from the lymphoid organs of a sheep immunized with melanoma cells from a fresh surgical specimen obtained from a third melanoma patient, HO. Some of these results[36] are illustrated in Figures 21 and 22. Of particular interest is the fact that RNA from the lymphoid organs of a sheep immunized with melanoma cells

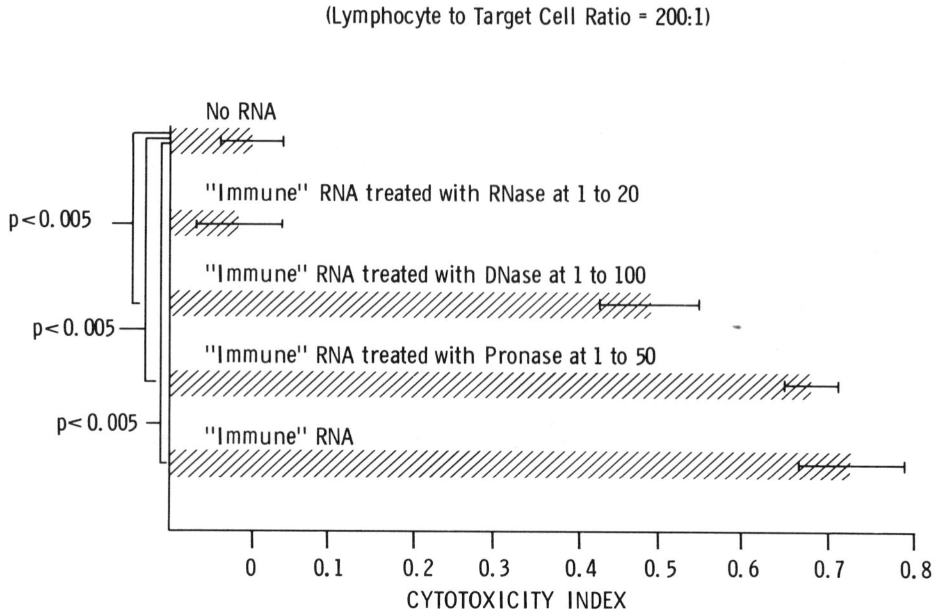

FIGURE 20. Effects of enzyme treatments on the activity of anti-BT-20 I-RNA. RNase was used at an enzyme-to-substrate ratio of 1 to 20. DNase and pronase were used at enzyme-to-substrate ratios of 1 to 100 and 1 to 50, respectively, based on the actual amount of DNA and protein in the I-RNA preparation.

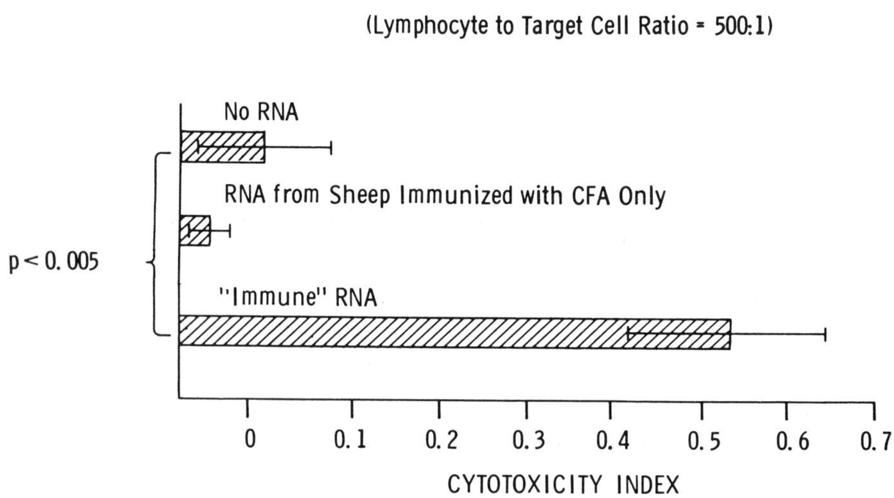

FIGURE 21. Immune cytolysis of RO-H melanoma cells mediated by xenogeneic I-RNA extracted from the lymphoid organs of a sheep immunized with melanoma cells from patient HO.

from one patient, HO, was effective in mediating cytotoxic immune responses against melanoma cells from two other patients, RO and ED.

A critical experiment was then performed using autologous lymphocytes obtained from the same patient, ED, from whom the ED-H cell line originated. These lymphocytes were incubated with sheep antimelanoma I-RNA and reacted with cells from this patient's own tumor in vitro. The patient's lymphocytes incubated without RNA served as a control. The results of this experiment are presented in Figure 23. After incubation without RNA, the patient's lymphocytes were slightly cytotoxic toward his tumor cells (CI=O.22

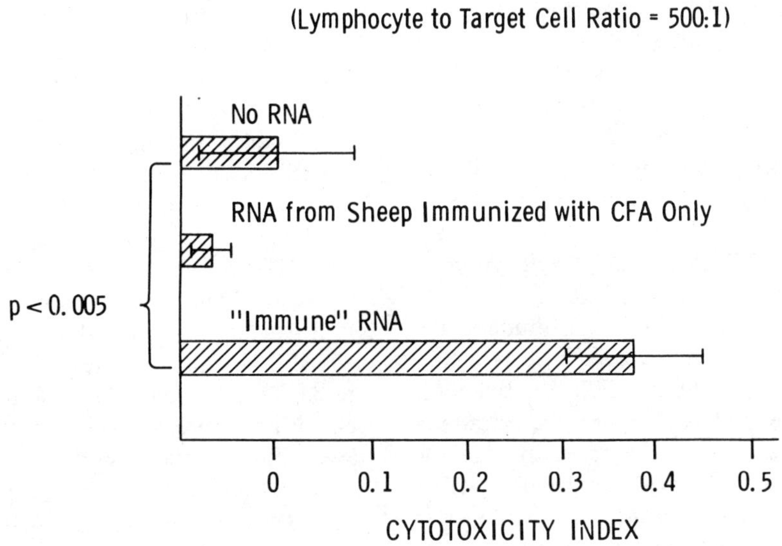

FIGURE 22. Immune cytolysis of ED-H melanoma cells mediated by xenogeneic I-RNA extracted from the lymphoid organs of a sheep immunized with melanoma cells from patient HO.

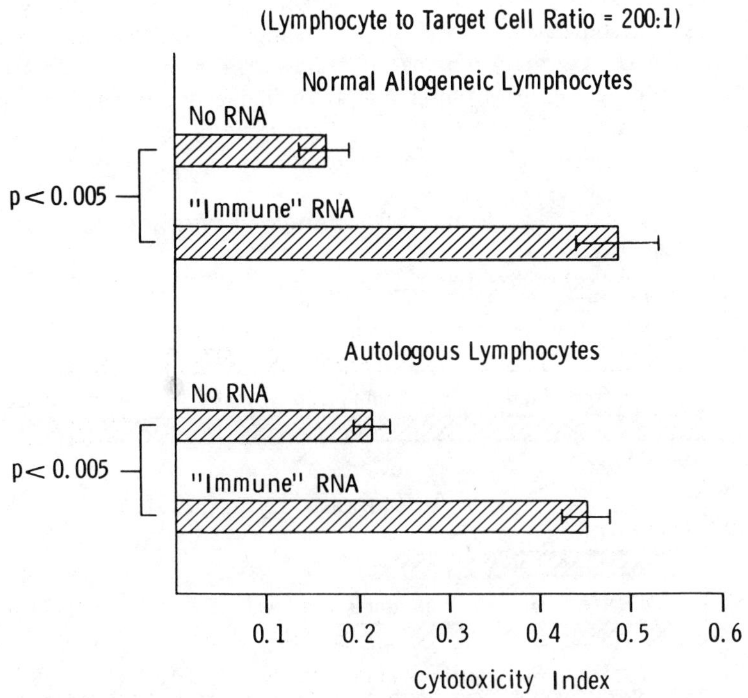

FIGURE 23. Immune cytolysis of ED-H melanoma cells by autologous lymphocytes and by allogeneic lymphocytes (from a normal donor) following incubation with xenogeneic I-RNA extracted from the lymphoid tissues of a sheep immunized with melanoma cells from patient HO. (From Pilch, Y. H., Fritze, D., Waldman, S. R., and Kearne, D. H., *Curr. Top. Microbiol. Immunol.,* 72, 157, 1975; *Curr. Probl. Surg.,* p. 26, February 1975. With permission.)

± 0.05). However, after the patient's lymphocytes were incubated with antimelanoma I-RNA, the cytotoxicity index of his lymphocytes more than doubled (CI=0.45 ± 0.02). This increase was significant at $p<0.005$. At the same time, lymphocytes from a normal volunteer were incubated with aliquots of the same preparation of I-RNA. The resulting CI was 0.49 ± 0.05. It can be seen that the effect of I-RNA on *autologous* lymphocytes (autologous with respect to the origin of the tumor target cells) was equal in magnitude to its effect on normal nonimmune allogeneic lymphocytes from a healthy volunteer. When autologous lymphocytes were used as effector cells, cytotoxic immune responses directed specifically against melanoma-associated antigens mediated by the I-RNA were almost certainly involved. Again, it is important to remember that the antimelanoma I-RNA was extracted from the lymphoid organs of a sheep immunized with melanoma cells from a *different* melanoma patient, suggesting that the mediation of immune responses to common melanoma specific antigens was involved.

To derive additional evidence that immune responses directed specifically against tumor-associated antigens could be mediated by I-RNA, a second experiment was performed which included additional controls. I-RNA was extracted from the lymphoid organs of either of two groups of guinea pigs. One group had been immunized with ED's melanoma cells (anti-ED-Mel RNA), while the second had been immunized with ED's autologous skin fibroblasts (anti-ED-Fib RNA). Each I-RNA preparation was incubated with normal allogeneic lymphocytes (from a normal healthy donor) or with ED's own lymphocytes. Following incubation, each lymphocyte sample was tested for its cytotoxic activity toward ED's melanoma target cells (ED-H cells). The results of this experiment are depicted in Figure 24. Note that when effector cells which were allogeneic with respect to the target cells were used, both anti-ED-Fib RNA and anti-Mel RNA were equally effective in mediating cytotoxic immune responses toward ED-H melanoma target cells. However, when autologous lymphocytes were used as effector cells, only anti-ED-Mel RNA was active, and anti-ED-Fib RNA did not lead to an increase in the cytotoxic activity of ED's lymphocytes for autologous melanoma cells. We interpret these results as follows. Both anti-ED-Mel RNA and anti-ED-Fib RNA induced allogeneic lymphocytes to effect immune responses directed against normal transplantation antigens (primarily HL-A antigens) as

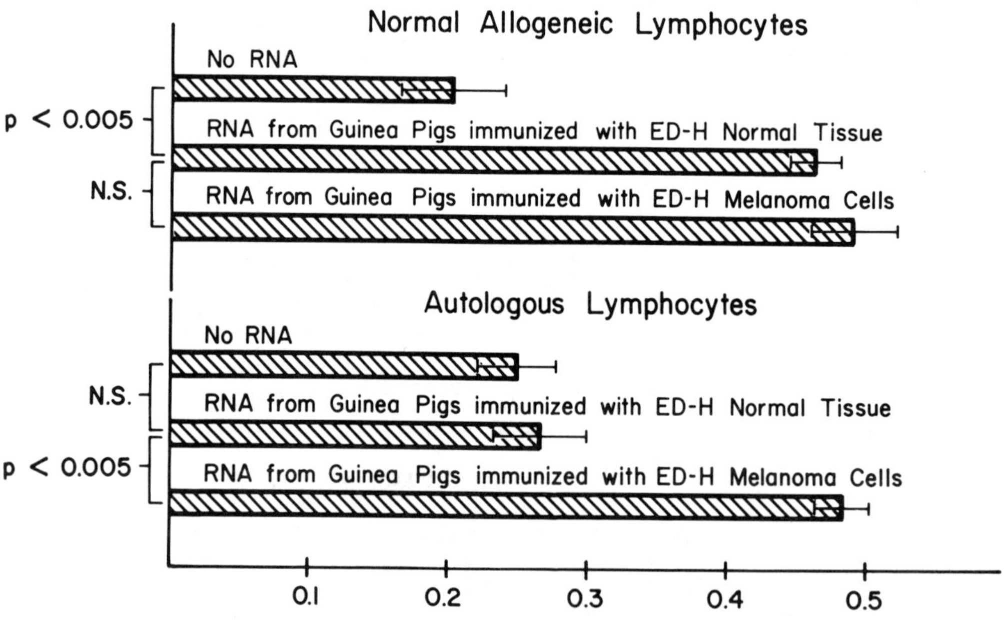

FIGURE 24. Immune cytolysis of ED-H melanoma cells by autologous lymphocytes and by allogeneic lymphocytes (from a normal donor) following incubation with xenogeneic I-RNA extracted from the lymphoid organs of guinea pigs immunized with ED-H melanoma cells or ED's autologous skin fibroblasts.

well as tumor-associated antigens (in the case of anti-ED-Mel RNA). When these I-RNA preparations were incubated with autologous lymphocytes, I-RNAs directed against normal transplantation antigens were recognized as "self" by the lymphocytes, and no immune responses against these antigens were elicited. Therefore, anti-ED-Fib RNA was inactive when incubated with autologous lymphocytes. Only I-RNAs directed against foreign antigens (i.e., tumor-associated antigens) were capable of eliciting antitumor immune responses from patient ED's lymphocytes. I-RNA extracted from lymphoid organs of animals immunized with ED's own fibroblasts, when incubated with ED's lymphocytes, failed to mediate cytotoxic immune responses against ED melanoma target cells, indicating a failure to elicit immune responses against "self" cell surface antigens.

In another experiment, aliquots of lymphocytes from a healthy volunteer were incubated with either xenogeneic I-RNA extracted from the lymphoid organs of guinea pigs immunized with melanoma tissue from patient CR or guinea pigs immunized with normal skin from melanoma patient CR (Figure 25). Both I-RNA against patient CR's normal tissue and I-RNA against CR's melanoma cells converted normal allogeneic lymphocytes to cells capable of mediating cytotoxic immune reactions toward melanoma target cells. However, when melanoma patient CR's own lymphocytes were incubated with aliquots of either of these same two I-RNA preparations, only I-RNA extracted from the lymphoid organs of guinea pigs immunized with CR's melanoma cells mediated cytotoxic immune reactions against melanoma target cells. I-RNA extracted from the lymphoid organs of guinea pigs immunized with CR's normal skin tissue, incubated with CR's lymphocytes, failed to mediate immune reactions against autocthonous melanoma target cells.

In another tumor system (Figure 26), it was found that only I-RNA extracted from the lymphoid organs of guinea pigs specifically immunized with breast cancer cells from patient ME, when incubated with ME's lymphocytes, mediated cytotoxic immune reactions against breast cancer target cells. RNA extracted from the lymphoid organs of guinea pigs immunized with breast cancer patient ME's normal muscle tissue was not active in converting ME's lymphocytes to "killer" cells. However, when either of these same

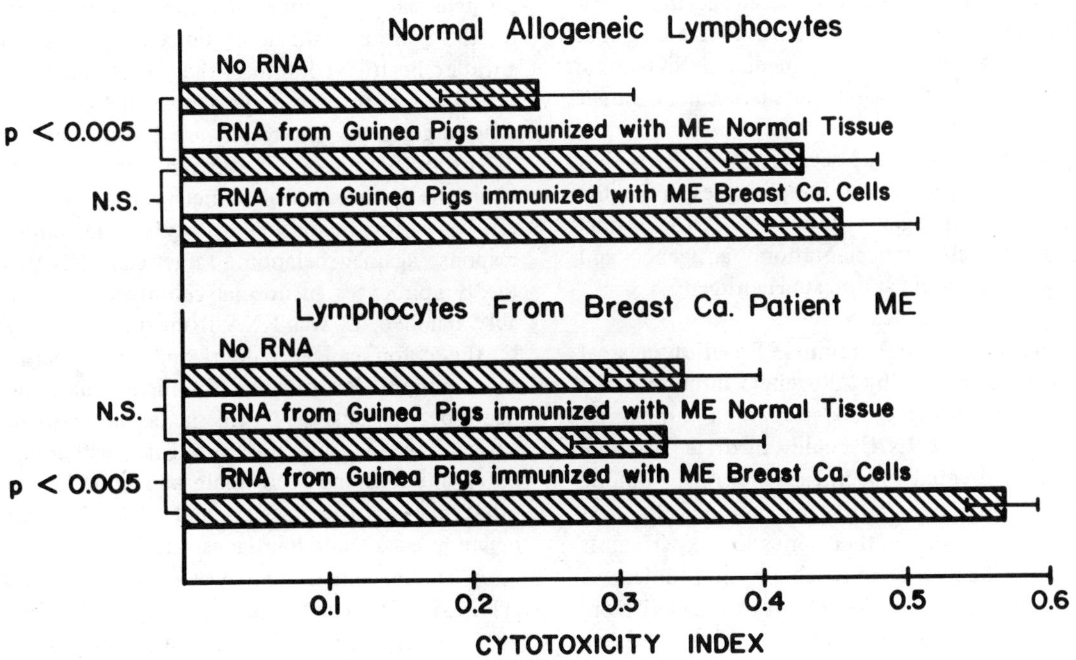

FIGURE 25. Immune cytolysis of ED-H melanoma cells by normal lymphocytes (from an allogeneic donor) and by lymphocytes from melanoma patient CR following incubation with xenogeneic I-RNA extracted from the lymphoid organs of guinea pigs immunized either with melanoma patient CR's melanoma cells or with CR's normal skin fibroblasts.

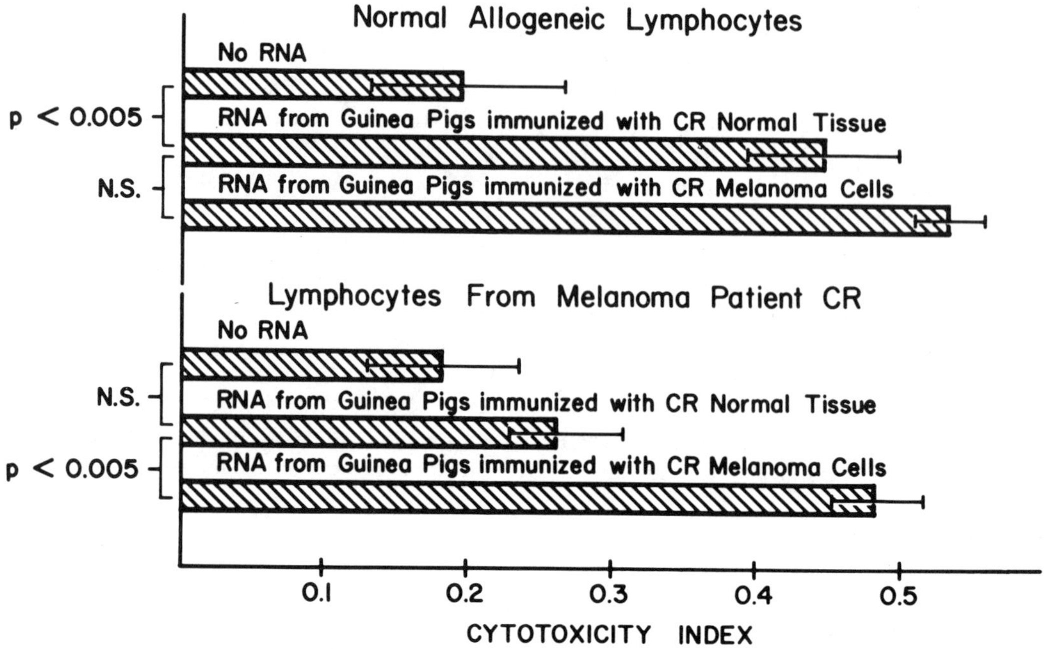

FIGURE 26. Immune cytolysis of BT-20 breast cancer cells by normal lymphocytes (from an allogeneic donor) and by lymphocytes from breast cancer patient ME following incubation with xenogeneic I-RNA extracted from the lymphoid organs of guinea pigs immunized either with patient ME's breast cancer cells or with ME's normal skin fibroblasts.

two RNAs was incubated with lymphocytes from a normal donor, I-RNA directed against ME's normal tissue, as well as I-RNA directed against ME's breast cancer cells, mediated cytotoxic immune responses against breast cancer target cells.

It is clear from these several experiments that human peripheral blood lymphocytes have the capacity to distinguish between I-RNA against normal, "self"-transplantation antigens and "foreign" (i.e., tumor-associated) antigens.

Mediation of Cytotoxic Immune Reactions against Human Cancer Cells by Allogeneic Immune RNA

Additional evidence to strengthen our earlier suspicion that I-RNA could mediate immune responses directed specifically against unique antigens associated with melanoma cells was derived from yet another series of experiments involving a different approach to the problem of specificity. Allogeneic I-RNA was extracted from the peripheral blood lymphocytes of each of five "cured" melanoma patients (collected on the continuous-flow blood-cell separator), one patient with colon carcinoma, one patient with hypernephroma, and five normal volunteers. Each I-RNA preparation was incubated with aliquots of a single sample of normal lymphocytes collected on the continuous-flow blood-cell separator from another healthy volunteer; then, each aliquot was tested for cytotoxic activity against ED-H melanoma target cells. The results are depicted in Figure 27. From these data it is apparent that only I-RNA from the lymphocytes of melanoma patients was effective in mediating a cytotoxic immune response against melanoma target cells. RNA from the lymphocytes of normal volunteers was invariably inactive, as was RNA from the lymphocytes of the colon cancer patient and the hypernephroma patient. Presumably, the melanoma patients had been sensitized to melanoma-associated antigens in vivo, but not to normal histocompatibility antigens (all were male and none had been multiply transfused), and I-RNA against melanoma-associated antigens was present.

A similar experiment was carried out using HT-29M colon cancer cells as target cells (Figure 28). I-RNAs extracted from the peripheral blood lymphocytes of three out of four colon cancer patients incubated with normal allogeneic human lymphocytes effected cytotoxic immune reactions against colon cancer target cells. RNAs extracted

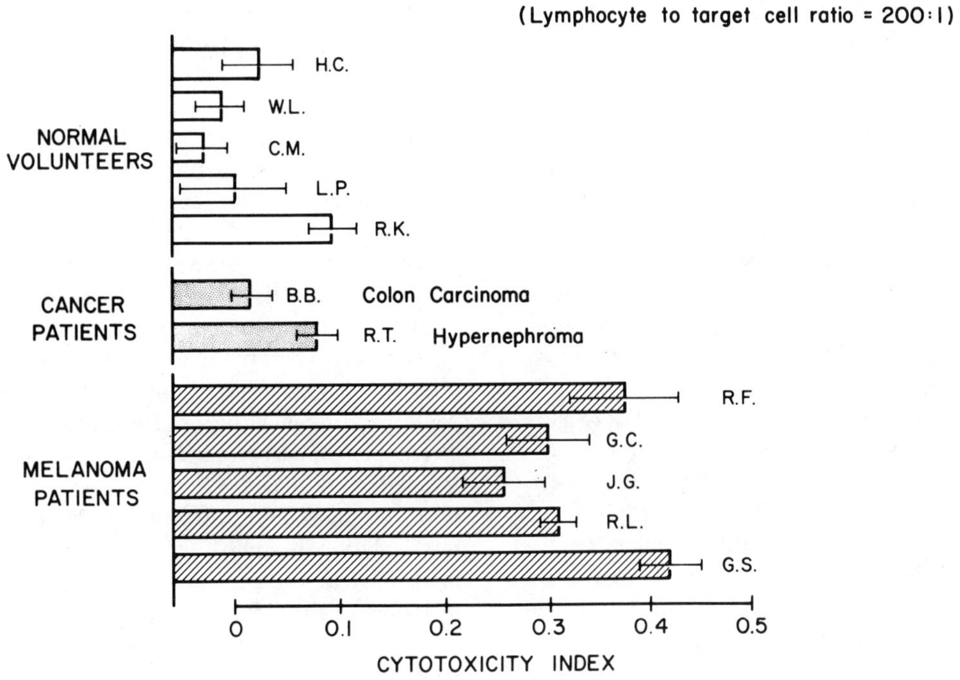

FIGURE 27. Immune cytolysis of ED-H melanoma cells mediated by allogeneic I-RNA extracted from the peripheral blood lymphocytes of "cured" melanoma patients.

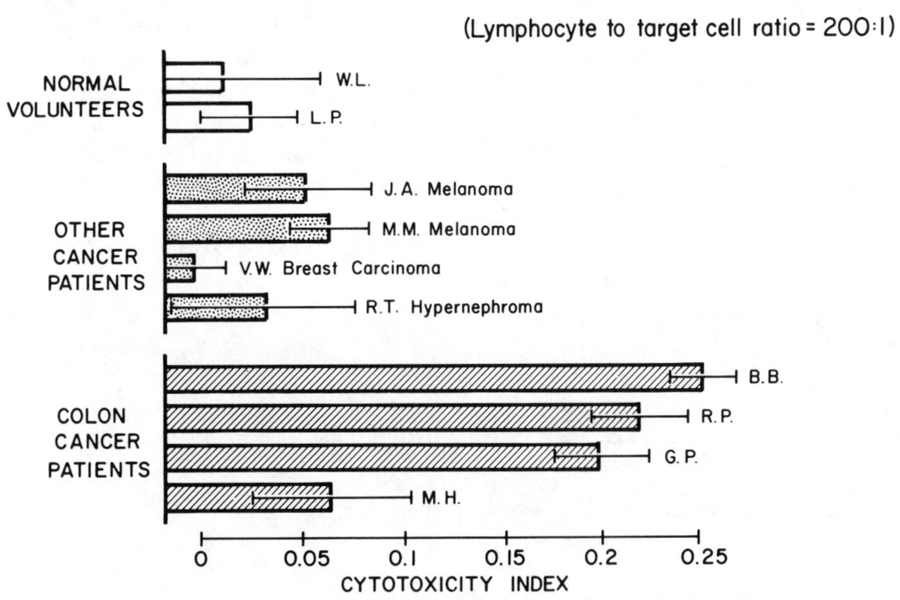

FIGURE 28. Immune cytolysis of colon cancer cells mediated by allogeneic I-RNA extracted from the peripheral blood lymphocytes of "cured" colon cancer patients.

from the peripheral blood lymphocytes of normal healthy volunteers were inactive. More importantly, I-RNAs extracted from lymphocytes of cancer patients with cancers other than colon cancer failed to elicit immune reactions directed against colon cancer target cells. These I-RNAs included two specimens from melanoma patients which had been found to be active in mediating immune responses against melanoma target cells. One of the I-RNAs from a colon cancer patient (B.B.) which was active against colon cancer target cells had previously been discovered to be inactive against melanoma cells.

From the results of another experiment (Figure 29), it can be seen that I-RNAs extracted from the peripheral blood lymphocytes of two breast carcinoma patients, incubated with normal lymphocytes from a healthy volunteer, mediated immune reactions against BT-20 breast cancer target cells. RNAs extracted from the peripheral blood lymphocytes of normal donors, or from patients with cancers other than breast carcinoma (including a colon cancer I-RNA active against colon cancer and two melanoma I-RNAs active against melanoma), failed to mediate significant cytotoxic immune responses against breast cancer target cells even though histocompatibility differences existed between the RNA donor and the "target" cells. RNA extracted from the lymphocytes of a third breast cancer patient was inactive in this assay. A sucrose density gradient profile indicated that this particular RNA was almost completely degraded and was probably inactive as a result.

None of the cancer patients from whose peripheral blood lymphocytes these I-RNAs were extracted had been multiply transfused, and none was multiparous (all male except for the breast cancer patients). It seems unlikely that they were sensitized to HL-A antigens or other normal human transplantation antigens in vivo, except as cross reactivity might exist between human histocompatibility antigens and ubiquitous (e.g., bacterial) foreign antigens. The patients, therefore, were presumedly sensitized only to tumor-associated antigens of their own tumors. We observed that only I-RNAs extracted from peripheral blood lymphocytes of melanoma patients mediated cytotoxic immune reactions against melanoma cells. Similarly, only I-RNAs extracted from the lymphocytes of colon cancer patients mediated cytotoxic immune responses against colon cancer target cells, and only I-RNAs from the lymphocytes of breast cancer patients mediated immune reactions against breast cancer target cells. In other words, cytotoxic immune reactions mediated by allogeneic I-RNA, extracted from the peripheral blood lymphocytes of cancer patients, were shown to be directed only against tumor target cells of the same histologic tumor

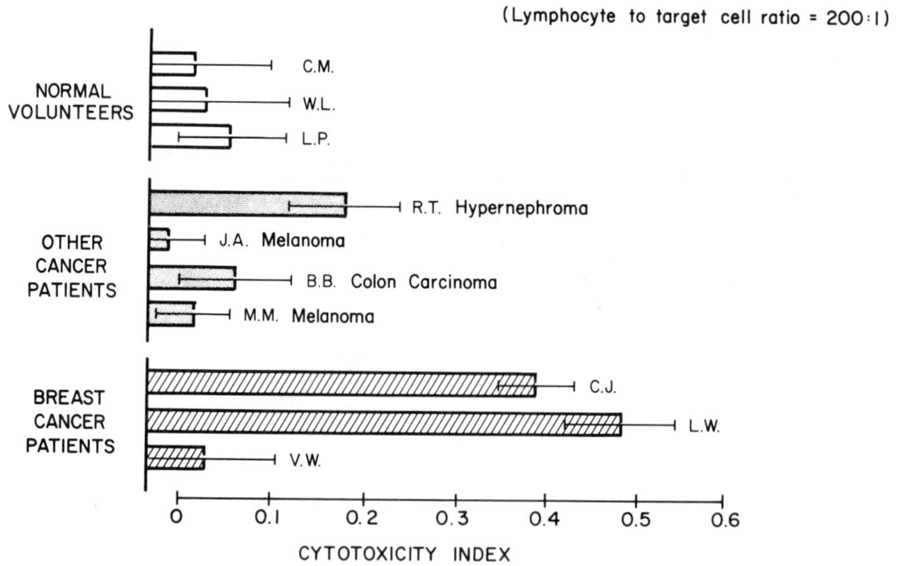

FIGURE 29. Immune cytolysis of BT-20 breast cancer cells mediated by allogeneic I-RNA extracted from the peripheral blood lymphocytes of "cured" breast cancer patients.

type as the I-RNA donor. It appears that allogeneic I-RNA extracted from the peripheral blood lymphocytes of cancer patients, under the conditions of our experiments, mediates cytotoxic immune reactions which are directed against common tumor-associated antigens specific for the tumor type of the RNA donor.

INITIAL CLINICAL TRIALS OF IMMUNOTHERAPY WITH IMMUNE RNA

Immune RNA offers several theoretical advantages over other methods of adoptive immunotherapy.

1. There is no need to give serum or plasma which has been shown in vitro in some instances to contain blocking factors which may facilitate tumor growth in vivo.
2. Since it is unnecessary to transfuse foreign leukocytes, sensitization of the recipient to foreign histocompatibility will not occur (HL-A antigens). Repeated transfusion of allogeneic leukocytes often results in significant toxic reactions of an allergic nature. I-RNA itself does not contain detectable transplantation antigens.
3. The danger of inducing graft-versus-host reaction in immunodeficient or immunosuppressed individuals is eliminated.
4. Since I-RNA itself is not a strong antigen, repeated injections of human or even animal RNA would not be expected to sensitize the recipient, cause secondary immune reactions, or result in immune elimination of subsequent doses of RNA.
5. I-RNA does not require the simultaneous administration of adjuvants.
6. I-RNA would probably be more effective as an immunogen than tumor antigens alone, assuming that good preparations of human tumor antigens eventually become available.
7. I-RNA might be effective despite certain types of host energy.

Two sources of I-RNA for the immunotherapy of cancer have been proposed, namely, allogeneic I-RNA derived from the lymphocytes of "cured" cancer patients and xenogeneic I-RNA derived from animals specifically immunized with neoplastic cells from the tumor-bearing patient or with tumor cells from another patient bearing a tumor of the same histologic type. Two methods by which I-RNA may be administered are proposed.

1. By parenteral injection into the patient.
2. By collective peripheral blood lymphocytes from the patient to be treated, exposing them in vitro with I-RNA, and then returning them intravenously to the patient.

Allogeneic I-RNA, like transfer factor, requires a human donor and, therefore, involves a number of inherent disadvantages.

1. The problem of donor selection.
2. The problem of donor availability.
3. The question of whether any human can provide lymphocytes which will yield an optimally immunoreactive extract.

Human tumors may be more immunogeneic in animals than they are in their host of origin. Also, will immunization of human donors be required or will "cured" cancer patients be adequate sources of I-RNA? Will repeated removal of lymphocytes from "cured" cancer patient donors result in sufficient immunosuppression to lead to an exacerbation of the donor's malignant disease? Due to these practical and theoretical problems with allogeneic I-RNA as an immunotherapeutic reagent, initial clinical trials have utilized I-RNA from animals rather than humans.

The successful interspecies transfer of cell-mediated immunity to human tumors in vitro (see above), provides a logical basis for the immunotherapy of human cancer with I-RNA extracted from the lymphoid tissues of animals specifically immunized with the type of human tumor to be treated. There are several obvious advantages to this form of immunotherapy. First, large quantities of I-RNA can be readily produced without dependence on human donors. Secondly, since histologically similar human tumors appear to contain cross-reactive tumor-associated antigens, many patients with the same tumor type might be treated with I-RNA from a sheep sensitized with a single patient's tumor.

We have recently undertaken a preliminary trial of immunotherapy of human cancer with xenogeneic I-RNA. At the time of this writing, 35 patients have been treated. The main intent of this study was to determine the safety and feasibility of immunotherapy with xenogeneic I-RNA and to determine the effects, if any, of this therapy on various immunologic parameters, both tumor specific and nonspecific. However, some interesting observations have been made regarding the effects of I-RNA therapy on certain antitumor immune responses assessed in vitro, and some indications of possible therapeutic benefit have been observed.

Two types of patients were studied: patients with grossly detectable and measureable metastatic disease and patients with "minimum residual disease." Patients in the latter category had no clinically detectable disease following surgical resection of all gross tumor, but had a greater than 50% likelihood of developing recurrent and/or metastatic disease within 24 months. The histologic types of tumors treated included malignant melanoma, hypernephroma, sarcoma, breast carcinoma, cholangiocarcinoma, and carcinoma of the stomach. The number of patients within each disease category and of each tumor type is given in Table 2. Of the three sarcoma patients, one had an alveolar soft part sarcoma, one a liposarcoma, and one an osteogenic sarcoma. Patients with gross disease were accepted for treatment only if:

1. They had failed all standard therapy (including standard chemotherapy).
2. No standard therapy of proven efficacy existed.
3. They had relatively stable disease and standard therapy could be interrupted or withheld for 8 weeks (the minimum duration of the study).

Tumor tissue obtained from surgical specimens (either freshly or obtained frozen in liquid nitrogen) was used. A thick suspension of viable tumor cells, emulsified in an equal volume of complete Freund's adjuvant, was injected intradermally into the extremities of sheep at weekly intervals for 3 consecutive weeks. Ten days after the last injection, the animals were sacrificed and I-RNA was extracted from their spleens and lymph nodes. The RNA was reprecipitated twice from solutions made 2 *M* with respect to potassium acetate, treated with pronase (to remove contaminating protein), and again reprecipitated from 2 *M* potassium acetate. The RNA was then dialyzed against sterile distilled water, sterilized by passage through a 0.22 μm Millipore filter, and lyophylized. It was resuspended in normal saline prior to injection. Aliquots were cultured to assure sterility.

Initially, the I-RNA was administered at a concentration of 1 mg RNA/ml. Later the concentration was increased to 2 mg RNA/ml. At higher concentrations the RNA was administered intradermally, in 0.1-ml aliquots, to the upper anterior chest near the axillae and lower abdomen near the groins. Usually the RNA was given at weekly intervals, rotating injection sites. Initially, the weekly dose of RNA was 1 mg. Later 2 mg/week was given, and most recently 4 mg/week has been the usual dose. A few patients have received up to 12 mg/week. Whenever possible, each patient received I-RNA from sheep injected with his own tumor tissue. Alternatively, I-RNA against allogeneic tumor tissue of the same histologic type was used. Treatment was given for 8 weeks, following which initial responses were evaluated. If progression of disease was noted, I-RNA therapy was discontinued and alternative therapy instituted.

No significant local or systemic toxicity has been noted with I-RNA therapy administered weekly for up to 18 months. Pain at the injection site was minimal. Local irritation, as evidenced by

TABLE 2
Number of Patients Treated with "Immune" RNA Grouped According to Tumor Type and Extent of Disease

Tumor types treated	Gross disease	Minimum residual disease	Total
Malignant melanoma	11	4	15
Hypernephroma	11	1	12
Sarcoma	3	0	3
Breast Carcinoma	0	2	2
GI carcinoma	1	1	2
Cholangiocarcinoma	1	0	1
Total	27	8	35

erythema, induration, or swelling, was absent or very minimal. No febrile reactions of any kind have occurred, and no allergic or anaphylactoid reactions have resulted.

The 35 patients treated with I-RNA ranged in age from 10 to 69. All were sensitized and skin tested with DNCB and a battery of four common skin test antigens (PPD, monilia, mumps, and streptokinase/streptodornase) prior to the initiation of immunotherapy; all were immunocompetent.

The duration of therapy to date ranged from 18 months to less than 2 months. Responses to therapy in patients with gross disease were evaluated by the following criteria.

1. Patients were considered "improved" if grossly detectable (and measurable) lesions objectively regressed and, following regression, did not increase in size for at least 2 months.
2. Patients were considered treatment "failures" if gross lesions significantly increased in area or new lesions appeared.
3. Patients were considered as having achieved "stability" if lesions which had been growing progressively prior to the initiation of therapy ceased all growth and remained stable for at least 2 months.
4. A few patients were classified as having achieved "possible benefit" when transitory regressions occurred or when prolonged stability was later followed by regrowth while on therapy.
5. Patients with minimum residual disease were considered "stable" as long as no recurrence or metastases developed.

Recurrence of the tumor was considered a treatment failure. When progression occurred, I–RNA therapy was discontinued.

The clinical results are summarized in Table 3. The first 26 patients are being reported, in detail, elsewhere.[37,38] In reviewing the results of this study to date, it must be remembered that clinical remissions were not expected at this state of our work, and any clinical responses reported must be considered anecdotal. In this initial Phase I trial, the objectives were

1. To establish the safety (or toxicity) of sheep I-RNA.
2. To evaluate dosage schedules and routes of administration.
3. To monitor any possible effects of I-RNA treatment on immunologic parameters, both tumor specific and nonspecific.

The optimum dosage and route and frequency of administration of I-RNA are not known. Certainly, we have not as yet approached a toxic dose of I-RNA.

In our previous animal experiments, it appeared desirable to administer I-RNA in a medium containing a strong ribonuclease inhibitor (the polyanion sodium dextran sulfate). However, as the

TABLE 3

Summary of Clinical Results

Tumor types	Improvement	Stability or possible improvement	Failure	Indeterminate	Total
Malignant melanoma					
Gross	0	4	6	1	11
Minimum residual		3	0	1	4
Hypernephroma					
Gross	3	4	2	3	12
Minimum residual		1	0	0	1
Sarcoma					
Gross	1	0	1	1	3
Breast carcinoma					
Minimum residual		1	0	0	1
Cholangiocarcinoma					
Gross	0	0	0	1	1
GI carcinoma					
Minimum residual		0	1	1	2
Total	4	13	10	8	35

safety of administration of dextran sulfate in man has not been determined as yet, we have not incorporated this ribonuclease inhibitor into our RNA preparations. Perhaps by so doing we might significantly increase the efficacy of I-RNA therapy.

Although treatment of lymphocytes in vitro with I-RNA has been shown to induce such lymphocytes to effect antitumor responses in vivo and in vitro, we have not as yet treated any patients by the intravenous infusion of the patient's lymphocytes preincubated in vitro with I-RNA. This reluctance has been due to the fear of inducing untoward allergic reactions related to small amounts of sheep protein which contaminate the RNA and might be expected to remain with the lymphocytes even after several washes. However, this method of utilizing I-RNA therapeutically may offer promise of greater efficacy.

Many of our patients had large tumor burdens and may not have been good candidates for immunotherapy of any kind. Perhaps (and this may be true for all types of immunotherapy) the primary usefulness of I-RNA will prove to be in the therapy of minimum residual disease. The number of patients with minimum residual disease treated in this manner is small. The follow-up period for most of these patients is very short, and we have no concomitantly generated control population. Therefore, it would be premature to make any conclusions from the results observed in these patients.

In a number of instances, lymphocytes obtained from patients prior to the onset of treatment were incubated, in vitro, with the I-RNA which they were to receive subsequently. After incubation of the lymphocytes with I-RNA for 20 min at 37°C, the lymphocytes were tested for cytotoxic effect in vitro on allogeneic tumor target cells of the same histologic type.

One example is presented in the experiment illustrated in Figure 30. Pretreatment lymphocytes from melanoma patient CA were tested against allogeneic melanoma cells (ED-H) before and after incubation with I-RNA extracted from the lymphoid organs of a sheep immunized with CA's own autologous tumor cells and after incubation with I-RNA extracted from the lymphoid organs of an unimmunized sheep. Following incubation with I-RNA from a sheep immunized with CA's tumor cells, the cytotoxicity of CA's lymphocytes for melanoma cells was increased significantly, while incubation with I-RNA from an unimmunized sheep was ineffective.

Another example is presented in the experiment illustrated in Figure 31. In this experiment, pretreatment lymphocytes from a gastric carcinoma patient were tested against allogeneic gastric carcinoma cells (RI-H) following (1) incubation without RNA, (2) incubation with

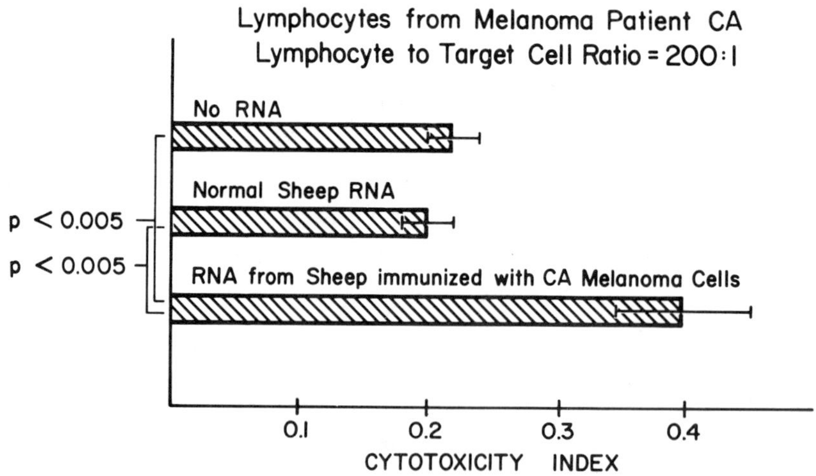

FIGURE 30. Cytotoxicity of pretreatment lymphocytes from melanoma patient CA for ED-H melanoma cells prior to and following incubation of I-RNA extracted from the lymphoid tissues of a sheep immunized with CA's own autologous tumor tissue and (as a control) I-RNA extracted from the lymphoid organs of a normal sheep.

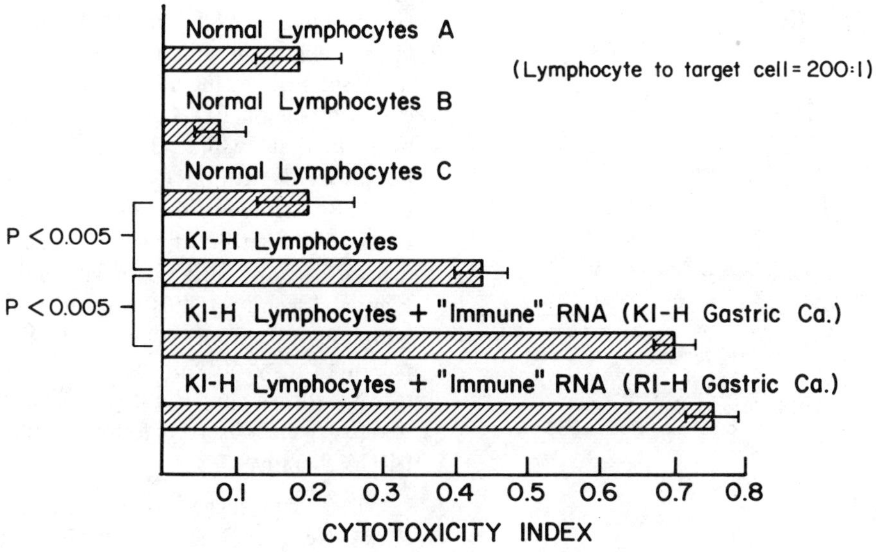

FIGURE 31. Cytotoxicity for gastric carcinoma (RI-H) target cells of lymphocytes from three normal donors and of pretreatment lymphocytes from gastric carcinoma patient KI incubated (1) with no RNA, (2) with RNA from a sheep immunized with KI's own tumor cells, and (3) with RNA from a sheep immunized with RI-H cells. (From Pilch, Y. H., Fritze, D., Waldman, S. R., and Kern, D. H., *Curr. Top. Microbiol. Immunol.*, 72, 157, 1975. With permission.)

I-RNA extracted from the lymphoid organs of a sheep immunized with KI's tumor cells, and (3) incubation with I-RNA extracted from the lymphoid organs of a sheep immunized with RI-H tumor cells. These are compared with the cytotoxic effects of lymphocytes from three different normal individuals which evidenced little toxic effect when tested against gastric carcinoma cells. Lymphocytes from gastric carcinoma patient KI, obtained prior to treatment, were significantly (but not strongly) cytotoxic for gastric carcinoma target cells. Following incubation with I-RNA from a sheep immunized with KI's tumor cells, the cytotoxicity of KI's lymphocytes for gastric carcinoma cells was greatly increased. The magnitude of this increase was the same as that observed when KI's pretreatment lymphocytes were incubated with I-RNA from a sheep immunized with the gastric carcinoma cells (RI-H) used as target cells in the assay.

In 14 of the 35 patients treated with I-RNA, peripheral blood lymphocytes, from serial bleedings obtained prior to and weekly during I-RNA therapy, were tested for cytotoxic effects against allogeneic target cells of the same histologic type. Of five patients in whom a significant increase in cytotoxic effects was noted, four evidenced stability or possible benefit in their clinical course. One was a treatment failure. Of eight patients in whom no change in lymphocyte-mediated cytotoxicity was noted, four evidenced stability or possible benefit, while four failed treatment. In one patient, a decrease in lymphocyte cytotoxic effect was noted, and this patient was a treatment failure. Table 4 presents a correlation between changes in lymphocyte-mediated cytotoxic responses assessed in vitro and clinical response. A possible correlation between increases in lymphocyte-mediated cytotoxicity, assessed in vitro, and clinical response is apparent.

TABLE 4

Correlation between Changes in Lymphocyte-mediated Cytotoxicity In Vitro and Clinical Response

Cytotoxicity	Stability or possible benefit	Failure	Total
Increase	4	1	5
No change	4	4	8
Decrease	0	1	1

The results of experiments performed in three of these patients are presented in detail in Figures 32 to 34. In Figure 32a the results of testing the

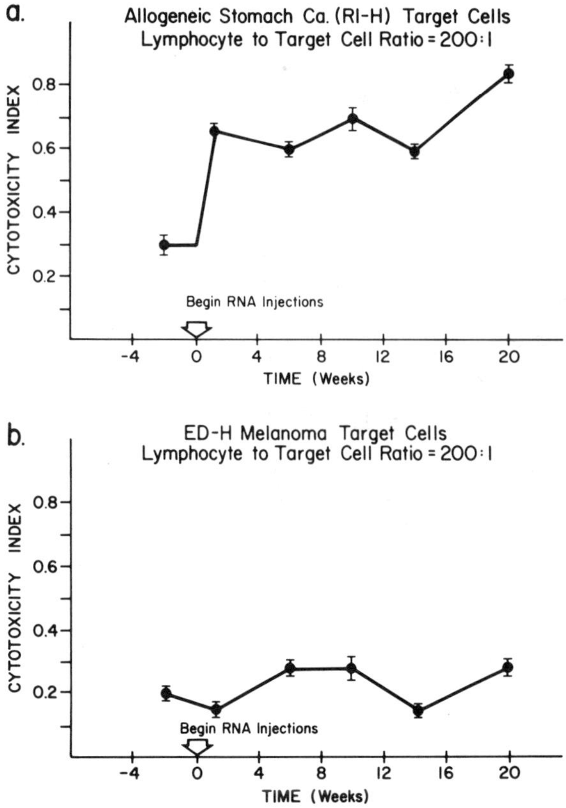

FIGURE 32. Cytotoxicity for RI-H gastric carcinoma target cells (Figure 32a) and for ED-H melanoma cells (Figure 32b) of lymphocytes from gastric carcinoma patient KI prior to and following the initiation of treatment with xenogeneic I-RNA extracted from the lymphoid organs of a sheep immunized with KI's own tissue. (From Pilch, Y. H., Fritze, D., Waldman, S. R., and Kern, D. H., *Curr. Top. Microbiol. Immunol.*, 72, 147, 1975. With permission.)

cytotoxicity of lymphocytes from gastric carcinoma patient KI on RI-H gastric carcinoma target cells is presented, while Figure 32b depicts the results of testing aliquots of the same lymphocyte samples on ED-H melanoma target cells. It is evident that while the cytotoxicity of KI's lymphocytes for RI–H cells increased substantially following the initiation of therapy with I-RNA, no significant cytotoxic activity was evidenced against ED-H melanoma target cells. Figures 33 and 34 illustrate the results of testing lymphocytes from two different melanoma patients for cytotoxic activity of ED-H melanoma target cells. Again, significant increases in cytotoxicity were observed following the initiation of I-RNA therapy.

These data suggest that treatment with I-RNA *does* effect at least certain immunologic parameters. This implies that these RNA preparations *are*, at least to some extent, immunologically active in man when administered in the dose and manner described above. These data also provide some evidence to suggest that the immune responses effected by I-RNA may, at least in part, be specific for the particular tumor type used to immunize the sheep from whose lymphoid tissues the I-RNA was extracted.

The results of the in vitro mictocytotoxicity tests performed in the lymphocytes of our patients do appear encouraging. Cytotoxicity responses were significantly increased after pretreatment lymphocytes were incubated with I-RNA. Moreover, in most of the patients tested, the cytotoxic

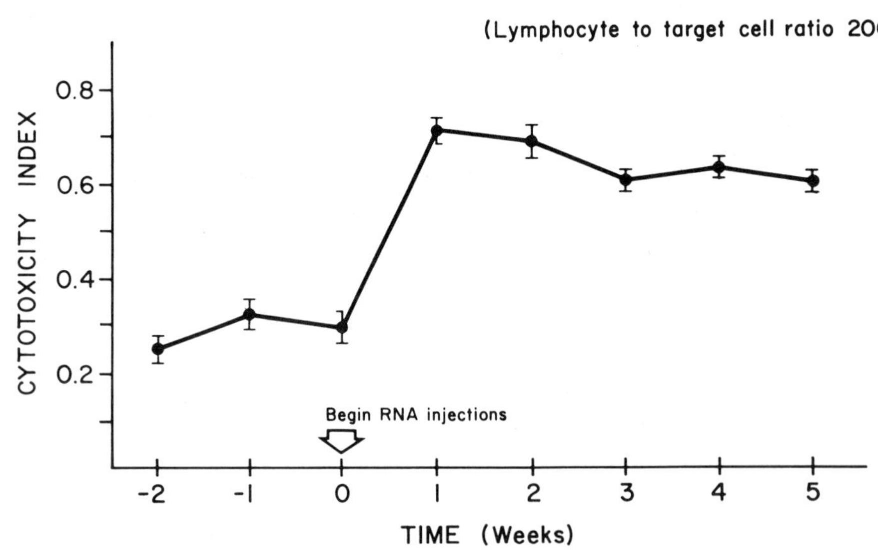

FIGURE 33. Cytotoxicity for ED-H melanoma cells of lymphocytes from melanoma patient SM prior to and following initiation of therapy with I-RNA extracted from the lymphoid organs of a sheep immunized with melanoma cells.

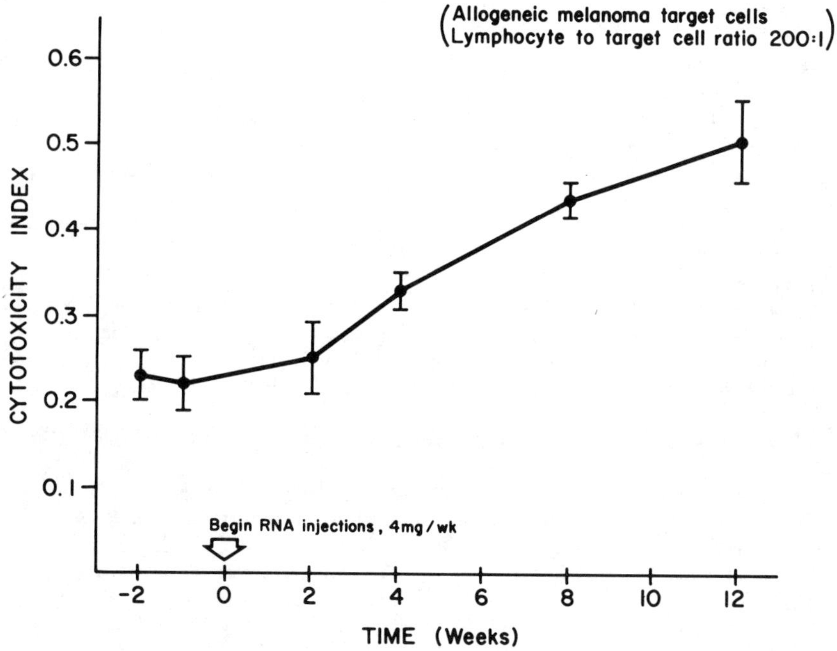

FIGURE 34. Cytotoxicity for ED-H melanoma cells of lymphocytes from melanoma patient VE prior to and following initiation of therapy with I-RNA extracted from the lymphoid organs of a sheep immunized with melanoma cells.

effects of peripheral blood lymphocytes for tumor cells increases substantially following I-RNA therapy, and we have some evidence that this increase is specific for the particular type of tumor used to immunize the RNA donor sheep.[39] It, therefore, appears that I-RNA treatment *does* effect host immune responses. We do not know if the nature or magnitude of this effect is relevant to the clinical course of human cancer or if it can result in therapeutic benefit. However, we have some evidence that these cytotoxic responses, assessed in vitro, may correlate with clinical responses to therapy.

Finally, it is clear that sheep I-RNA, when prepared as described above and administered intradermally in the doses and schedules described, is completely free of significant local and systemic toxicity and is very well tolerated. I-RNA offers several practical and theoretical advantages as a new modality for the immunotherapy of cancer. Much further study will be required to determine its potential immunotherapeutic efficacy.

REFERENCES

1. **Mannick, J. A. and Egdahl, R. H.,** Transformation of nonimmune lymph node cells to state of transplantation immunity by RNA. A preliminary report, *Ann. Surg.,* 156, 356, 1962.
2. **Mannick, J. A. and Egdahl, R. H.,** Transfer of heightened immunity to skin homografts by lymphoid RNA, *J. Clin. Invest.,* 43, 2166, 1964.
3. **Mannick, J. A.,** Fractionation of RNA capable of transferring transplantation immunity, in *Nucleic Acids in Immunology,* Plescia, O.J. and Braun, W., Eds., Springer-Verlag, New York, 1968, 547.
4. **Sabbadini, E. and Sehon, A. H.,** Transfer of transplantation immunity with RNA from lymphoid organs of immunized animals, in *Advance in Transplantation,* Dausset, J., Hamburger, J., and Mathe, G., Eds., Williams and Wilkins, Baltimore, 1968, 55.
5. **Sabbadini, E. and Sehon, A. H.,** Induction of accelerated graft rejection and enhanced graft-versus-host reaction with RNA from lymphoid organs of animals immunized with allogeneic tissues, in *Nucleic Acids in Immunology,* Plescia, O. J. and Braun, W., Eds., Springer-Verlag, New York, 1968, 560.
6. **Sabbadini, E. and Sehon, A. H.,** Acceleration of allograft rejection induced by RNA from sensitized donors, *Int. Arch. Allergy Appl. Immunol.,* 32, 55, 1967.
7. **Ramming, K. P. and Pilch. Y. H.,** Transfer of transplantation immunity by ribonucleic acid, *Transplantation,* 7, 296, 1969.
8. **Wilson, D. B. and Wecker, E. E.,** Quantitative studies on the behavior of sensitized lymphoid cells in vitro. III. Conversion of "normal" lymphoid cells to an immunologically active status with RNA derived from isologous lymphoid tissues of specifically immunized rats, *J. Immunol.,* 97, 512, 1966.
9. **Bondevik, H. and Mannick, J. A.,** RNA-mediated transfer of lymphocyte vs. target cell activity, *Proc. Soc. Exp. Biol.,* 129, 264, 1968.
10. **Alexander, P., Delorme, E. J., Hamilton, L. D., and Hall, J. G.,** Stimulation of anti-tumor activity of the host with RNA from immune lymphocytes, in *Nucleic Acids in Immunology,* Plescia, O. J. and Braun, W., Eds., Springer-Verlag, New York, 1968, 527.
11. **Alexander, P., Delorme, E. J., Hamilton, L. D., and Hall, J. G.,** Effect of nucleic acids from immune lymphocytes on rat sarcomata, *Nature,* 213, 569, 1967.
12. **Rigby, P. G.,** Prolongation of survival of tumor-bearing animals by transfer of "Immune" RNA with DEAE dextran, *Nature,* 221, 968, 1969.
13. **Londner, M. V., Morini, J. C., Font, M. T., and Rabasa, S. L.,** RNA-induced immunity against a rat sarcoma, *Experientia,* 24, 598, 1968.
14. **Kennedy, C. T. C., Cater, D. B., and Hartveit, F.,** Protection of C3H mice against BP-8 tumor by RNA extracted from lymph nodes and spleens of specifically sensitized mice, *Acta Pathol. Microbiol. Scand. (A),* 77, 196, 1969.
15. **Ramming, K. P. and Pilch, Y. H.,** Mediation of immunity to tumor isografts in mice by heterologous ribonucleic acid, *Science,* 168, 492, 1970.
16. **Ramming, K. P. and Pilch, Y. H.,** Transfer of tumor-specific immunity with RNA: Inhibition of growth of murine tumor isografts, *J. Natl. Cancer Inst.,* 46, 735, 1971.
17. **Pilch, Y. H. and Ramming, K. P.,** Transfer of tumor immunity with ribonucleic acid, *Cancer,* 26, 630, 1970.
18. **Deckers, P. J. and Pilch, Y. H.,** Mediation of immunity to tumor-specific transplantation antigens by RNA inhibition of isograft growth in rats, *Cancer Res.,* 32, 839, 1972.
19. **Friedman, H.,** Specific restoration of agglutinin formation in x-irradiated rabbits by nucleoprotein fractions from immune donors, *Ann. N.Y. Acad. Sci.,* 114, 444, 1964.
20. **Friedman, H.,** The nature of immunogenic RNA-antigen complexes in immune and tolerant mice, in *Nucleic Acids in Immunology,* Plescia, O. J. and Braun, W., Eds., Springer-Verlag, New York, 1968, 505.
21. **Bach, M. K.,** The inhibition of deoxyribonucleotidyl transferase RNase and DNase by sodium polyethene sulfonic acid. Effect of the molecular weight of the inhibitor, *Biochim. Biophys. Acta,* 91, 619, 1964.
22. **Fellig, J. and Wiley, C. E.,** The inhibition of pancreatic ribonuclease by anionic polymers, *Arch. Biochem. Biophys.,* 85, 313, 1959.
23. **Mora, P. T. and Young, B. G.,** Reversible inhibition of enzymes by interaction with synthetic polysaccharide macroanions, *Arch. Biochem Biophys.,* 82, 6, 1959.
24. **Deckers, P. J. and Pilch, Y. H.,** Transfer of immunity to tumor isografts by the systemic administration of xenogeneic "Immune" RNA, *Nat. New Biol.,* 231, 181, 1971.
25. **Pilch, Y. H., de Kernion, J. B., and Deckers, P. J.,** Transfer of tumor immunity with RNA, *Isr. J. Med. Sci.,* 7, 246, 1971.
26. **Granger, G. A. and Weiser, R. S.,** Howograft target cells: specific destruction in vitro by contact interaction with immune macrophages, *Science,* 145, 1427, 1964.
27. **Morton, D. L., Goldman, L., and Wood, D. A.,** Tumor-specific antigenicity of methylcholanthrene (MCA) and dibenzanthracene (DBA) induced sarcomas of guinea pigs, *Fed. Proc.,* 25, 372, 1965.
28. **Holmes, E. C., Kahan, B. D., and Morton, D. L.,** Soluble tumor-specific transplantation antigens from methylcholanthrene-induced guinea pig sarcomas, *Cancer,* 25, 373, 1970.

29. **Ramming, K. P. and Pilch, Y. H.,** Transfer of tumor-specific immunity with RNA: demonstration by immune cytolysis of tumor cells in vitro, *J. Natl. Cancer Inst.,* 45, 543, 1970.
30. **Kern, D. H. and Pilch, Y. H.,** Immune cytolysis of urine tumor cells mediated by xenogeneic "Immune" RNA, *Int. J. Cancer,* 13, 679, 1974.
31. **Kern, D. H., Drogemuller, C. R., and Pilch, Y. H.,** Immune cytolysis of rat tumor cells mediated by syngeneic "Immune" RNA, *J. Natl. Cancer Inst.,* 52, 299, 1974.
32. **Kern, D. H., de Kernion, J. B., and Pilch, Y. H.,** Intracellular localization of anti-tumor "Immune" RNA, *Cell. Immunol.,* in press.
33. **Veltman, L. L., Kern, D. H., and Pilch, Y. H.,** Immune cytolysis of human tumor cells mediated by xenogeneic "Immune" RNA, *Cell. Immunol.,* 13, 367, 1974.
34. **Pilch, Y. H., Veltman, L. L., and Kern, D. H.,** Immune cytolysis of human tumor cells mediated by xenogeneic "Immune" RNA: implications for immunotherapy, *Surgery,* 76, 23, 1974.
35. **Pilch, Y. H., Veltman, L. L., and Kern, D. H.,** Immune cytolysis of human tumor cells mediated by xenogeneic "Immune" RNA, *Arch. Surg.,* 109, 30, 1974.
36. **Pilch, Y. H., Fritz, D., and Kern, D. H.,** Immune cytolysis of human melanoma cells mediated by Immune RNA, *Behring Inst. Mitt.,* 56, 184, 1975.
37. **Pilch, Y. H., deKernion, J. B., Skinner, D. G., Ramming, K. P., Schick, P. M., Fritze, D., Brower, P., and Kern, D. H.,** Immunotherapy of cancer with "Immune" RNA: A preliminary report, *Am. J. Surg.,* in press.
38. **Skinner, D. G., deKernion, J. B., and Pilch, Y. H.,** Advanced renal cell carcinoma: Treatment with xenogeneic immune ribonucleic acid (RNA) and appropriate surgery, *J. Urol.,* in press.
39. **Pilch, Y. H. and Ramming, K. P.,** unpublished data.

INDEX

E

F

G

H

I

K

R

S

T